Live Variola Virus

CONSIDERATIONS FOR CONTINUING RESEARCH

Committee on the Assessment of
Future Scientific Needs for Live Variola Virus
Board on Global Health

Ann M. Arvin and Deepali M. Patel, *Editors*

INSTITUTE OF MEDICINE
OF THE NATIONAL ACADEMIES

THE NATIONAL ACADEMIES PRESS
Washington, D.C.
www.nap.edu

THE NATIONAL ACADEMIES PRESS 500 Fifth Street, N.W. Washington, DC 20001

NOTICE: The project that is the subject of this report was approved by the Governing Board of the National Research Council, whose members are drawn from the councils of the National Academy of Sciences, the National Academy of Engineering, and the Institute of Medicine. The members of the committee responsible for the report were chosen for their special competences and with regard for appropriate balance.

This study was supported by Contract No. BOGH-H-07-05-A between the National Academy of Sciences and the Centers for Disease Control and Prevention. Funds originated with both the Centers for Disease Control and Prevention and the Office of the Biomedical Advanced Research and Development Authority of the U.S. Department of Health and Human Services. Any opinions, findings, conclusions, or recommendations expressed in this publication are those of the author(s) and do not necessarily reflect the view of the organizations or agencies that provided support for this project.

International Standard Book Number-13: 978-0-309-13690-7
International Standard Book Number-10: 0-309-13690-3

Additional copies of this report are available from the National Academies Press, 500 Fifth Street, N.W., Lockbox 285, Washington, DC 20055; (800) 624-6242 or (202) 334-3313 (in the Washington metropolitan area); Internet http://www.nap.edu.

For more information about the Institute of Medicine, visit the IOM home page at: **www.iom.edu.**

Printed in the United States of America

Front cover: Reprinted with Permission by Luke Jerram, Artist in Residence at FACT, Liverpool for European Capital of Culture 2008.

The serpent has been a symbol of long life, healing, and knowledge among almost all cultures and religions since the beginning of recorded history. The serpent adopted as a logotype by the Institute of Medicine is a relief carving from ancient Greece, now held by the Staatliche Museen in Berlin.

Suggested citation: IOM (Institute of Medicine). 2009. *Live Variola Virus: Considerations for Continuing Research*. Washington, DC: The National Academies Press.

"Knowing is not enough; we must apply.
Willing is not enough; we must do."
—Goethe

INSTITUTE OF MEDICINE
OF THE NATIONAL ACADEMIES

Advising the Nation. Improving Health.

THE NATIONAL ACADEMIES

Advisers to the Nation on Science, Engineering, and Medicine

The **National Academy of Sciences** is a private, nonprofit, self-perpetuating society of distinguished scholars engaged in scientific and engineering research, dedicated to the furtherance of science and technology and to their use for the general welfare. Upon the authority of the charter granted to it by the Congress in 1863, the Academy has a mandate that requires it to advise the federal government on scientific and technical matters. Dr. Ralph J. Cicerone is president of the National Academy of Sciences.

The **National Academy of Engineering** was established in 1964, under the charter of the National Academy of Sciences, as a parallel organization of outstanding engineers. It is autonomous in its administration and in the selection of its members, sharing with the National Academy of Sciences the responsibility for advising the federal government. The National Academy of Engineering also sponsors engineering programs aimed at meeting national needs, encourages education and research, and recognizes the superior achievements of engineers. Dr. Charles M. Vest is president of the National Academy of Engineering.

The **Institute of Medicine** was established in 1970 by the National Academy of Sciences to secure the services of eminent members of appropriate professions in the examination of policy matters pertaining to the health of the public. The Institute acts under the responsibility given to the National Academy of Sciences by its congressional charter to be an adviser to the federal government and, upon its own initiative, to identify issues of medical care, research, and education. Dr. Harvey V. Fineberg is president of the Institute of Medicine.

The **National Research Council** was organized by the National Academy of Sciences in 1916 to associate the broad community of science and technology with the Academy's purposes of furthering knowledge and advising the federal government. Functioning in accordance with general policies determined by the Academy, the Council has become the principal operating agency of both the National Academy of Sciences and the National Academy of Engineering in providing services to the government, the public, and the scientific and engineering communities. The Council is administered jointly by both Academies and the Institute of Medicine. Dr. Ralph J. Cicerone and Dr. Charles M. Vest are chair and vice chair, respectively, of the National Research Council.

www.national-academies.org

COMMITTEE ON THE ASSESSMENT OF FUTURE SCIENTIFIC NEEDS FOR LIVE VARIOLA VIRUS

ANN M. ARVIN (*Chair*), Lucile Packard Professor of Pediatrics and Microbiology and Immunology, Stanford University School of Medicine

DONALD BURKE, Dean, Graduate School of Public Health, University of Pittsburgh

LIANG DENG, Assistant Member, Dermatology Service, Department of Medicine, Memorial Sloan-Kettering Cancer Center, New York

DIANE E. GRIFFIN, Professor and Chair in Molecular Microbiology and Immunology, Johns Hopkins School of Public Health

STEPHEN OSTROFF, Director, Pennsylvania Department of Health, Bureau of Epidemiology, Harrisburg

PETER A. PATRIARCA, Senior Clinical Consultant, Biologics Consulting Group, Inc., Bethesda

CLARENCE J. PETERS, Professor, Department of Microbiology and Immunology and Pathology, University of Texas Medical Branch at Galveston

DAVID A. RELMAN, Thomas C. and Joan M. Merigan Professor of Medicine, and of Microbiology and Immunology, Stanford University; Chief, Infectious Diseases, VA Palo Alto Health Care System

MARIA SALVATO, Professor, Institute of Human Virology, University of Maryland School of Medicine

MARK SLIFKA, Associate Scientist, Vaccine and Gene Therapy Institute, Oregon Health and Sciences University

HOLLY TAYLOR, Assistant Professor, Department of Health Policy and Management, Johns Hopkins Bloomberg School of Public Health

PAULA TRAKTMAN, Professor, Department of Microbiology and Molecular Genetics, Medical College of Wisconsin

DAVID ULAETO, Principal Scientist, Department of Biomedical Sciences, Defense Science and Technology Laboratory, Salisbury, UK

Study Staff

DEEPALI PATEL, Program Officer

KATHERINE MCCLURE, Senior Program Associate

ALLISON BRANTLEY, Senior Program Assistant

PATRICK KELLEY, Director, Board on Global Health

CARLY LANGLAIS, Consultant

JULIE WILTSHIRE, Financial Associate

Reviewers

This report has been reviewed in draft form by individuals chosen for their diverse perspectives and technical expertise, in accordance with procedures approved by the National Research Council's Report Review Committee. The purpose of this independent review is to provide candid and critical comments that will assist the institution in making its published report as sound as possible and to ensure that the report meets institutional standards for objectivity, evidence, and responsiveness to the study charge. The review comments and draft manuscript remain confidential to protect the integrity of the deliberative process. We wish to thank the following individuals for their review of this report:

Kenneth I. Berns, University of Florida Genetics Institute
R. Mark L. Buller, Department of Molecular Microbiology and Immunology, St. Louis University Health Sciences Center
R. Alta Charo, University of Wisconsin Law School
Lawrence Corey, Vaccine and Infectious Disease Institute, Fred Hutchinson Cancer Research Center and Virology Department, University of Washington
Kathryn M. Edwards, Department of Pediatrics, Vanderbilt University School of Medicine
William M. Egan, PharmaNet Consulting
Joseph J. Esposito, JGL Tech-consult
Thomas Inglesby, Center for Biosecurity, University of Pittsburgh Medical Center

W. Ian Lipkin, Center for Infection and Immunity, Mailman School of Public Health, Columbia University
D. Grant McFadden, Department of Molecular Genetics and Microbiology, University of Florida College of Medicine
Philip K. Russell, Major General (retired) USA, Professor Emeritus, Johns Hopkins School of Public Health
Ronald K. St. John, St. John Public Heath Consulting International Inc., Ontario
Robert Swanepoel, Special Pathogens Unit, National Institute for Communicable Diseases, South Africa
Richard Whitley, Department of Pediatrics and Microbiology, University of Alabama

Although the reviewers listed above have provided many constructive comments and suggestions, they were not asked to endorse the conclusions or recommendations nor did they see the final draft of the report before its release. The review of this report was overseen by **Dr. Floyd E. Bloom,** Professor Emeritus, Department of Molecular and Integrative, Neuroscience, The Scripps Research Institute, and **Dr. Adel A.F. Mahmoud,** Woodrow Wilson School and Department of Molecular Biology, Princeton University. Appointed by the National Research Council and Institute of Medicine, they were responsible for making certain that an independent examination of this report was carried out in accordance with institutional procedures and that all review comments were carefully considered. Responsibility for the final content of this report rests entirely with the authoring committee and the institution.

Preface

Smallpox is caused by variola virus, which modern genomics reveals to be very closely related to vaccinia and other orthopoxviruses. These similarities mean that vaccination protects against smallpox, as Edward Jenner observed in the late eighteenth century. Transmission of variola in the human population, its only natural host, was halted in the twentieth century through the dedicated efforts of public health workers and volunteers on every continent. Smallpox was declared eradicated by the World Health Assembly in 1980, a moment that is recognized as one of the most important achievements of mankind.

In the course of the smallpox eradication campaign, laboratories that held variola virus isolates or clinical specimens that might contain infectious virus destroyed the materials or sent them to one of two repositories that were established in the United States and Russia under the direction of the World Health Organization (WHO). Since then, the World Health Assembly has debated whether the live variola virus stocks or other potentially infectious materials collected at these two repositories should be destroyed. The World Health Assembly will take up the issue of the retention or destruction of live variola virus stocks in 2010. Therefore, the Institute of Medicine (IOM) was asked by the Centers for Disease Control and Prevention and the Office of the Biomedical Advanced Research and Development Authority of the U.S. Department of Health and Human Services to form a committee that would conduct a study on the continued use of live variola virus for research and public health purposes.

This committee's work follows upon that of an earlier IOM committee that was formed in 1998 to assess future scientific needs for live variola virus. The previous committee's deliberations were undertaken in preparation for the 1999 meeting of the World Health Assembly, which was to address what should be done with the variola stocks being maintained at the Centers for Disease Control and Prevention in the United States and at the State Centre for Research of Virology and Biotechnology (VECTOR) in Russia. In 1999, the IOM committee released its report, *Assessment of Future Scientific Needs for Live Variola Virus*, in which it offered a number of consensus conclusions. These conclusions focused particularly on the uses of live variola virus for developing medical countermeasures against smallpox. At that time, the World Health Assembly chose to defer its decision on destroying the variola stocks until such research could be conducted in the future. In the interim, WHO has been responsible for overseeing all research involving live variola virus, which has been allowed only under the highest level of biosafety containment at the two WHO Collaborating Centers for Smallpox and Other Poxviruses.

In this context, the present committee undertook an examination of the scientific needs for live variola virus, based on a critical assessment of the research that has been reported in the decade since the first IOM report was published. It is important to note that this committee, like its 1999 counterpart, was not asked to consider whether the stocks should be retained or destroyed, nor was it charged with evaluating the risks of an accidental or intentional release of variola virus.

To address its charge, the committee organized two workshops for public presentations by leading experts, including those with experience from the smallpox eradication campaign, investigators who have done basic and applied research with live variola virus, and others who addressed topics relevant to the committee's task. The committee members also reviewed the scientific literature related to variola and other poxviruses in their specific areas of expertise and summarized their findings for the committee. The committee formally requested from VECTOR information on a range of aspects of their variola-related research. Professor Ilya G. Drozdov, Director-General, Head, WHOCC for Orthopoxvirus Diagnosis and Repository for Variola Virus Strains and DNA, provided extensive information on accomplishments at VECTOR over the past decade that have been cited in the report. The workshops and the analysis of the literature served as the background for extensive discussions and the development of conclusions by the committee in the course of three convened meetings.

Through its deliberations, the committee reached consensus on the circumstances under which live variola virus would be essential and others under which it would be useful for research or public health purposes, as presented in this report. The report is intended to provide information about what has been accomplished since the 1999 IOM committee's assessment and to offer recommendations to basic and clinical researchers, policy makers, and the public regarding the scientific uses of live variola virus in the contemporary context.

Ann M. Arvin, *Chair*
Committee on the Assessment of Future Scientific Needs for Live Variola Virus

Acknowledgments

The committee and the Institute of Medicine wish to express their appreciation to the many individuals who took the time to share their expert knowledge and perspectives with the committee in the course of its work and to the organizations that provided necessary information.

The committee would like to thank the following individuals for their presentations at the public workshops: Kenneth Berns, Paul Chaplin, Inger Damon, Joseph Esposito, Donald Henderson, Lisa Hensley, Sally Hojvat, Dennis Hruby, Peter Jahrling, Kevin Karem, Daniel Lavanchy, Grant McFadden, Michael Merchlinsky, Hermann Meyer, Bernard Moss, Gerald Parker, Anne Rimoin, Kate Rubins, Barbara Styrt, and Richard Whitley. The committee would like to thank Professor Ilya G. Drozdov, Director-General, Head, WHOCC for Orthopoxvirus Diagnosis and Repository for Variola Virus Strains and DNA, for his detailed response to its inquiries concerning the variola-associated work that has been conducted at VECTOR.

The committee's work could not have been completed without the skilled professional support of the Institute of Medicine staff. We are grateful to Allison Brantley, Patrick Kelley, Carly Langlais, Katherine McClure, and Deepali Patel.

Contents

Summary

Smallpox was a devastating disease that decimated human populations for centuries, and its eradication in 1980 was a monumental achievement for the global health community. Since then the remaining known stocks of its causative agent, variola virus, a member of the *Orthopoxvirus* genus, have been contained in two World Health Organization (WHO)–approved repositories—in the United States at the Centers for Disease Control and Prevention in Atlanta, Georgia; and in Russia at the Research Institute for Viral Preparations in Moscow, transferred in 1994 to the State Centre for Research of Virology and Biotechnology (VECTOR) in Novosibirsk.

In 1999, the World Health Assembly (WHA) debated the issue of destroying these remaining stocks. Arguments were presented on the need to retain the live virus for use in additional important research, and the decision to destroy the virus was deferred until this research could be completed. In that same year, the Institute of Medicine (IOM) convened a consensus committee to explore scientific needs for the live virus. The IOM committee reached a number of conclusions, focused in particular on the need to develop medical countermeasures for smallpox and the role of the live virus in meeting this need (see Box S-1).

Ten years have passed since that committee conducted its analysis, and the scientific, political, and regulatory environments have changed. Technological advances have led to breakthroughs in drug development and genomic analysis. The 2001 anthrax attacks in the United States demonstrated the feasibility of using a biological agent as a weapon of terror and the need for better detection and control methods for such threats.

BOX S-1
Conclusions from the 1999 Institute of Medicine Report

1. Genomic sequencing and limited study of variola surface proteins derived from geographically dispersed specimens is an essential foundation for important future work. Such research could be carried out now, and could require a delay in the destruction of known stocks, but would not necessitate their indefinite retention.
2. The most compelling reason for long-term retention of live variola virus stocks is their essential role in the identification and development of antiviral agents for use in anticipation of a large outbreak of smallpox. It must be emphasized that if the search for antiviral agents with activity against live variola virus were to be continued, additional public resources would be needed.
3. Adequate stocks of smallpox vaccine must be maintained if research is to be conducted on variola virus or if maintenance of a smallpox vaccination program is required. Live variola virus would be necessary if certain approaches to the development of novel types of smallpox vaccine were pursued.
4. If further development of procedures for the environmental detection of variola virus or for diagnostic purposes were to be pursued, more extensive knowledge of the genome variability, predicted protein sequences, virion surface structure, and functionality of variola virus from widely dispersed geographic sources would be needed.
5. The existence of animal models would greatly assist the development and testing of antiviral agents and vaccines, as well as studies of variola pathogenesis. Such a program could be carried out only with live variola virus.
6. Live or replication-defective variola virus would be needed if studies of variola pathogenesis were to be undertaken to provide information about the response of the human immune system.
7. Variola virus proteins have potential as reagents in studies of human immunology. Live variola virus would be needed for this purpose only until sufficient variola isolates had been cloned and sequenced.

SOURCE: IOM, 1999, pp. 82–85.

As a result, licensing requirements for medical countermeasures for use in such circumstances have become more comprehensive. In this new climate, the IOM was once again tasked to consider scientific needs for live variola virus.

With the body of knowledge that has accumulated in the past 10 years has come new insight into the fundamental biology of variola. In particular, understanding of the virus's unique adaptation to its sole host—humans—has implications for learning more about the human host response to viral infection. A deeper understanding of the life cycle of variola and its ability to subvert immune defense has provided, and will continue to provide, potential novel targets for antiviral and vaccine design. Genomic advances have led to the sequencing of at least 48 geographically diverse isolates of variola, as well as the theoretical potential for resynthesis of whole genomes, broadening the knowledge base on all orthopoxviruses and setting the stage for the possible development of surrogates for live variola virus in research. And finally, increased attention to the threat of bioweapons of terror has led to further refinement of the regulatory pathway to approved use and licensure of medical countermeasures.

The current IOM committee was charged with revisiting the question of scientific needs for live variola virus; like its predecessor, this committee was not asked to consider the issue of retention versus destruction of the existing stocks of the virus. In addressing its charge (see Box S-2), the committee made a concerted effort to perform a comprehensive assessment that encompassed research in both the United States and the rest of the world.

BOX S-2
Charge to the Committee

An *ad hoc* committee of the Institute of Medicine shall conduct a study on the continued use of live variola virus stocks for research and public health purposes. In follow-on to the IOM's 1999 report, *Assessment of Future Scientific Needs for Live Variola Virus,* an IOM committee will perform a comprehensive evaluation of the research and development work recommended in that report and completed to date, and consider what unmet needs still exist that require the use of live variola virus. The conclusions and recommendations will inform policy discussions in the United States and within the world community regarding the continued need to retain the official stocks of live variola virus for research purposes, and would provide a major review of completed, ongoing and planned research activities that should be undertaken.

The committee shall specifically consider and offer recommendations pertinent to the utility of live variola virus in addressing potential unmet requirements including:

- Advanced development through licensure and post-licensure of antivirals for use in treatment of variola virus infections.
- Advanced development through licensure and post-licensure of new, safe and effective vaccine(s).
- Development through licensure and post-licensure of less-reactogenic vaccines.
- Development of approved protein-based diagnostics which can be used in field situations or diagnostics which have sources of error distinct from those of nucleic acid-based diagnostics.
- Improved pathogenesis data to drive therapeutic discovery.

It considered both practical and theoretical research while making no judgment on the infrastructure for or financial feasibility of either.

In formulating its conclusions and recommendations, the committee drew a distinction between those uses for live variola virus which are important and essential and those which are useful but not essential. The committee considered the development of medical countermeasures against this deadly pathogen—including therapeutics, vaccines, and diagnostic tools—to be an important and essential need because of the potential for an accidental or deliberate release.

CONCLUSIONS

The committee evaluated the scientific need for live variola virus in four areas: development of therapeutics, development of vaccines, genomic analysis, and discovery research. The committee's conclusions in each of these areas are presented below and summarized in Table S-1.

Development of Therapeutics

Currently, no therapeutic to treat smallpox infection exists. The previous IOM committee determined that the development of therapeutics was the most immediate need requiring retention of the live variola virus stocks. Since then, a number of candidate drugs have been developed, one of which has been approved for compassionate use. Work still remains on the full development and licensure of these drugs, as well as on the discovery and development of other therapeutics with different targets and adverse effect profiles. In addition, research into other therapies, such as those that enhance or modulate immune response, could result in alternative or adjunctive treatments for smallpox. Finally, based on experience with other pathogens, the threat of drug resistance remains a real possibility, and drug

TABLE S-1 Overview of Essential Versus Useful Scientific Needs for Live Variola Virus

Need	Requires Use of Live Virus	Does Not Require Live Virus
Essential	• Development of therapeutics and assessment of resistance • Development of vaccines that do not manifest a take	• Development of first- and second-generation vaccines that produce a take • Development of methods for detection and diagnosis
Useful	• Functional genomics-based research • Discovery research	• Variola genome sequence analysis

designs and regimens that could reduce the emergence of resistance need to be investigated.

The committee concludes that, for both scientific and regulatory reasons, the final developmental stages leading to licensure of smallpox therapeutics cannot occur without the use of live variola virus. Furthermore, although the regulatory environment may change, the scientific reasons will remain. Therapeutic agents need to be evaluated against a representative panel of variola strains to reduce the possibility that some strains might be naturally resistant.

Development of Vaccines

Thirty years after the eradication of smallpox, a majority of the world's population is no longer immune to the disease. Despite the known effectiveness of currently licensed smallpox vaccines, the small risk of adverse events contraindicates their use in specific subpopulations, such as the immunocompromised. In response, research into less reactogenic and safer vaccines has been conducted, resulting in the licensure of a second-generation vaccinia vaccine. Research is also continuing on third-generation vaccines, which have the potential to protect vulnerable populations. However, second-generation vaccines use the same strain of vaccinia virus as the first-generation vaccines; validation with live variola virus has not been necessary; and efficacy can be assessed through the manifestation of "take"—a cutaneous lesion that forms at the site of inoculation. Nonreplicating and subunit third-generation vaccines do not present a take; therefore, an estimate of their potential efficacy requires vaccination and challenge with live variola virus in animal models, as well as testing of the immune responses of human vaccine recipients with methods that show activity against the live virus.

The committee concludes that the current development and licensure pathway for first- and second-generation vaccinia vaccines that produce a "take" does not require use of the live variola virus. Use of the live virus will be necessary, however, for the development and licensure of any vaccine that does not manifest such a cutaneous lesion at the site of inoculation.

Development of Methods for Detection and Diagnosis

Contemporary nucleic acid-based methods for viral detection have been shown to identify variola virus genes directly, and multiplex polymerase chain reaction (PCR) assays differentiate variola from other poxviruses and unrelated viruses, such as varicella-zoster virus, that may cause similar clinical signs. Protein-based assays have not been pursued as extensively as PCR assays; however, these assays can be tested using variola proteins made in expression vectors. Limited information has been published about the performance of any methods for environmental sampling to detect variola. The primary barrier to development of these methods is a lack of development incentives and of a market for products that would allow rapid field detection and diagnosis.

The committee concludes that live variola virus is not required for further development of detection and diagnostic methods. Virus materials such as DNA and proteins would suffice for this purpose.

Genomic Analysis

Progress in sequencing variola strains has revealed some of the genetic variability of variola—in particular, similarities and differences among genes in various regions—although much genetic information remains to be discovered. In addition, the biological consequences of sequence differences have not been well explored. Beyond the genome, analysis of functional pathways and expressed proteins would yield even deeper understanding.

The committee concludes that live variola virus is not needed for variola genome sequence analysis, as long as specimens containing viral DNA of adequate quantity and quality are available. Live variola virus would be needed for functional genomics-based experimental approaches.

Discovery Research

Variola virus can be useful for understanding human physiology and immunology because it has the capacity to overwhelm the host in a way that few viral pathogens do. Through studies in nonhuman primates, some progress has been made in understanding how variola virus modulates the functions of host cells for its benefit and how infection with the virus progresses in the host. While there is no immediate need for this type of research, it could result in knowledge that might one day lead to the discovery of new drugs or vaccines. In particular, better understanding of the genomic variability among variola strains, of the differences and similarities among orthopoxviruses, and of the host response to variola (and other orthopoxvirus) infection could elucidate how best to enhance the ability to counter a smallpox outbreak or infection.

The committee concludes that discovery research to gain greater understanding of human physiology and immunology, while not essential, would require use of the live variola virus and might ultimately support efforts to discover and evaluate therapeutics and vaccines. Further, research with live variola virus and research with variola proteins could lead to discoveries with broader implications for human health.

RECOMMENDATIONS

In the context of the above conclusions, the committee recognizes additional research that would enhance understanding of variola virus. Gaps remain in knowledge about the virus and its interaction with its host that could be critical in identifying potential targets for drug and vaccine discovery. In particular, better understanding of the diversity and variability of variola strains would result in more refined diagnostics and more effective therapeutics. Genome sequencing could close existing knowledge gaps by illuminating differences in molecular mechanisms of infection and response.

The committee recommends that WHO authorize the complete genome sequencing of all remaining variola strains, with the aim of understanding the patterns and extent of sequence variation and the relationships of these patterns to disease severity. This activity would be carried out at CDC, and ideally at VECTOR as well.

Similarly, a better understanding of variola pathogenesis would enhance the development of therapeutics and vaccines. Because smallpox is no longer naturally occurring, the closest approximation to human infection would involve a nonhuman primate. A more precise nonhuman primate model is essential for correct characterization of the efficacy of new therapeutics and vaccines. It is important to optimize approaches to infecting nonhuman primates so as to best recapitulate variola pathogenesis as it occurs in the human host, for example, by testing aerosol or intratracheal delivery as well as intravenous inoculation.

The committee recommends that a comprehensive evaluation of the work done to date on the nonhuman primate model of variola pathogenesis be undertaken by CDC, in conjunction with an expert panel knowledgeable about poxviruses and animal models of viral infection. The objective would be to identify ways in which the predictive value of the model for testing therapeutics and vaccines might be improved.

Finally, functional genomic tools, which are used to evaluate interactions between a replicating virus and the host cell, should be applied using a few representative variola strains in a number of representative differentiated human cell types. The purpose of this research would be to identify novel targets for therapeutics and to design third-generation vaccines.

The committee recommends that WHO explore the use of functional genomics approaches to improve understanding of variola pathogenesis and advance the development of novel strategies for therapeutic intervention.

REFERENCE

IOM (Institute of Medicine). 1999. *Assessment of future scientific needs for live variola virus.* Washington, DC: National Academy Press.

1

Introduction

Throughout history, humankind has been plagued by a number of deadly diseases. Smallpox, perhaps the most devastating of these, has always been greatly feared (Morens et al., 2008). The earliest description of smallpox as a distinct clinical syndrome emerged in fourth-century CE China, but earlier records hint at its presence in Europe before then (Damon, 2006). Its causative agent, variola virus, has adapted in unique ways to its only known host species—humans.

After centuries of recurring smallpox epidemics that swept through human populations worldwide, impacted the course of history, and killed more than 500 million people, Edward Jenner reported in 1798 that inoculation with related viruses, obtained from either cows or horses, conferred immunity to smallpox (Jenner, 1798). More than 150 years later, with this critical tool in hand, the World Health Organization (WHO) in 1959 embarked on an ambitious plan to control and eventually eradicate the disease. In 1977, the last known naturally occurring case of smallpox was recorded in Somalia, and the following year an accidental laboratory-associated infection became the last known case of the disease. In 1979, a commission of health experts certified that natural transmission of smallpox had ceased, and WHO endorsed the declaration a year later in 1980. This final eradication of smallpox represents a monumental event in the history of medicine and public health, and smallpox remains the only intentionally eradicated disease of humankind (Fenner et al., 1998; Tucker, 2001).

Since the eradication of smallpox, the global public health community, acting through the World Health Assembly (WHA), has debated the issue of whether stocks of the live variola virus should be retained. In 1983, the number of centers authorized to house and perform research with the live virus was limited to two—the Centers for Disease Control and Prevention (CDC) in the United States and the Research Institute for Viral Preparations in Moscow, Russia. In 1994, the Russian stocks were transferred to the State Centre for Research of Virology and Biotechnology (VECTOR) in Novosibirsk (Fenner et al., 1998; Tucker, 2001).

In 1996, the WHO ad hoc Committee on Orthopoxvirus Infections recommended final destruction of the live variola virus stocks at both research centers, and WHA subsequently set the termination date for 1999. However, the decision to destroy the virus was postponed that year in light of increasing public health and biosecurity concerns, and was ultimately deferred by WHA to assess the potential for continued scientific needs for the live virus (Smallpox Preservation Advisable, 1999; WHO, 1999). In that same year, the Institute of Medicine (IOM) released a consensus report identifying key areas for further scientific research that required the use of live variola virus (IOM, 1999). The conclusions from that report are presented in Box 1-1.

Ten years have passed since the 1999 IOM report was issued, and much has since transpired that is relevant to the question of the utility of research using live variola virus. These developments include advances in science and biotechnology, incidents involving bioterrorism, increased investment in research and development on countermeasures, changes in the regulatory approval process, and the reinstitution of smallpox vaccination among civilian and military populations in the United States. Since 1999, WHO has convened a standing Advisory Committee on Variola Virus Research, which has met annually since 1999 and which monitors the state of research in key areas at the two WHO Collaborating Centers for Smallpox and Other Poxviruses. (The reports of the committee's meetings are available through the WHO website.) This committee reports its findings at the annual meetings of WHA, which is the ultimate decision-making body of WHO; these meetings are attended by delegations from all of the WHO member states. WHA has agreed to revisit the issue of variola virus destruction in 2010. In this context, it is important to re-assess the conclusions of the 1999 IOM report and to review the progress that has been made over the last decade.

OVERVIEW OF SCIENTIFIC NEEDS FOR LIVE VARIOLA VIRUS

Despite the successful eradication of smallpox 30 years ago, concerns remain about the potential for its reemergence. While a natural outbreak appears unlikely, the threat of intentional reintroduction or accidental release exists. In addition, the emergence of human disease due to monkeypox (another member of the *Orthopoxvirus* genus), including its 2003

BOX 1-1
Conclusions from the 1999 Institute of Medicine Report

In 1998–1999, the IOM convened a committee to evaluate the scientific needs for continued retention of live variola virus. The committee identified one short-term need and six long-term needs. Specifically, the committee drew the following conclusions:

- Genomic sequencing and limited study of variola surface proteins derived from geographically dispersed specimens is an essential foundation for important future work. Such research could be carried out now, and could require a delay in the destruction of known stocks, but would not necessitate their indefinite retention.
- The most compelling reason for long-term retention of live variola virus stocks is their essential role in the identification and development of antiviral agents for use in anticipation of a large outbreak of smallpox. It must be emphasized that if the search for antiviral agents with activity against live variola virus were to be continued, additional public resources would be needed.
- Adequate stocks of smallpox vaccine must be maintained if research is to be conducted on variola virus or if maintenance of a smallpox vaccination program is required. Live variola virus would be necessary if certain approaches to the development of novel types of smallpox vaccine were pursued.
- If further development of procedures for the environmental detection of variola virus or for diagnostic purposes were to be pursued, more extensive knowledge of the genome variability, predicted protein sequences, virion surface structure, and functionality of variola virus from widely dispersed geographic sources would be needed.
- The existence of animal models would greatly assist the development and testing of antiviral agents and vaccines, as well as studies of variola pathogenesis. Such a program could be carried out only with live variola virus.
- Live or replication-defective variola virus would be needed if studies of variola pathogenesis were to be undertaken to provide information about the response of the human immune system.
- Variola virus proteins have potential as reagents in studies of human immunology. Live variola virus would be needed for this purpose only until sufficient variola isolates had been cloned and sequenced.

SOURCE: IOM, 1999, pp. 82–85.

introduction into the western hemisphere, highlights the importance of research into prevention of orthopoxvirus infection and disease (Reed et al., 2004; Sale et al., 2006; Rimoin et al., 2007; Dubois and Slifka, 2008).

At the same time, tools to control potential smallpox outbreaks remain imperfect. While smallpox vaccines based on cross-protection provided by vaccinia-induced immunity are available and are known to be effective from extensive historical experience, there are still concerns regarding their safety. An estimated 40 percent of those vaccinated with the original and newer, second-generation vaccinia vaccines experience mild to severe adverse reactions (McCurdy et al., 2004). The development of third-generation vaccines with the potential to have a much improved safety profile is currently under way. Additionally, no U.S. Food and Drug Administration (FDA)-approved therapeutics or validated, rapid, point-of-care diagnostics for smallpox are available. The lack of these tools would be an obstacle in the event of a future outbreak of smallpox or other orthopox.

Advances made in the fields of molecular biology and genetics in the past 10 years could provide important tools to improve understanding of the structure of variola virus and the functions of its gene products. Additionally, given variola's specificity for the human species, studying live variola virus in cultured human cells in vitro—including both cells that are targets for initial infection and spread and specialized cells that mediate the immune response—holds the potential to yield new insights into the antiviral mechanisms of host cells and the biology of the human immune system. These findings could provide valuable information not only in the context of controlling DNA viruses, but also as a means of understanding basic inflammatory pathways that can protect or damage the host. Such observations could be extended in appropriate animal models of variola pathogenesis. Complete chemical resynthesis of the variola genome and subsequent production of viable intact virions is now scientifically plausible and technically feasible. This newly emerging and rapidly evolving capability has profound implications for both the future threat posed by smallpox and the future development of smallpox countermeasures.

Box 1-2 summarizes features of the contemporary context in which the scientific needs for live variola virus must be assessed.

CURRENT STATUS OF VARIOLA VIRUS AND MATERIALS

Because of the biohazard posed by live variola virus, rigorous precautionary measures are essential, including strict regulation of the type of facility that is approved for storing and conducting experiments with the virus. Research with live variola virus must be conducted in laboratories with the highest safety and security rating, designated biosafety level 4 (BSL-4) containment facilities. Although other BSL-4 facilities exist, only the two noted above in the United States and Russia are authorized by WHO to perform research with live variola virus under international agreement (WHO, 2008).

WHO oversees all scientific research with live variola virus, and to ensure the safety of researchers and the security of the virus stocks, periodically conducts inspections of the authorized research facilities. In addition

BOX 1-2
Contemporary Context for Assessment of the Scientific Needs for Live Variola Virus

Smallpox as a bioweapon. Smallpox's virulence makes it an obvious candidate for use as a bioweapon. Historical anecdotes, while not confirmed, suggest that contaminated materials could be used to spread smallpox in target populations. Both the United States and the Soviet Union have engaged in research aimed at weaponizing smallpox.

Monkeypox outbreaks. First recognized in humans in 1970, monkeypox is endemic in central Africa. Periodic outbreaks have occurred in the Democratic Republic of Congo, with a case fatality rate of 1–10 percent. In addition, the introduction of monkeypox in the United States in 2003 demonstrated the continuing threat of orthopoxvirus outbreaks.

Immunologically naïve populations. Routine smallpox vaccination ceased in 1980, and earlier in some countries. Almost half the world's population is currently immunologically naïve to the disease. At the same time, the rise of diseases such as HIV/AIDS that weaken the immune system, as well as the prevalence of atopic dermatitis, would make resumption of routine vaccination difficult.

Lack of proper countermeasures. There are today no licensed therapeutics for the treatment of smallpox, and currently licensed vaccines, while effective, are contraindicated for immunocompromised individuals.

Resynthesis of the variola genome. Technological advances have led to new breakthroughs, including the complete sequencing of multiple strains of variola virus. The ability to resynthesize viral genomes is well established and may be possible for variola virus.

to handling of the live virus, work with the genomic components of the virus is tightly regulated by WHO. Specifically, laboratories other than CDC and VECTOR cannot possess more than 20 percent of the variola genome at any time (WHO, 1994).

From its most recent (November 2008) meeting, the WHO Advisory Committee on Variola Virus Research reports that access to the BSL-4 laboratories at CDC and VECTOR remains highly controlled and regulated; security procedures are reviewed by WHO, and in the United States by the U.S. Select Agent Program. CDC has also reported on an expansion of its BSL-4 facilities, with another laboratory scheduled to be operational in 2009. Since November 2006, the long-term inventory of variola virus materials at CDC has remained at 451, and genomes from

45 of the 70 working stock isolates have been sequenced. Withdrawals have been made to support WHO-approved projects. Since 2007, a total of 200 nonviable or duplicate working stocks at VECTOR have been destroyed, reducing its collection of variola stocks to 691 registered vials (WHO, 2008).

REGULATIONS AND OTHER GUIDANCE PERTAINING TO COUNTERMEASURES FOR SMALLPOX

Since the IOM's 1999 report was issued, a number of regulations and other guidance have been promulgated in the United States to guide and facilitate the development and licensure of additional countermeasures for the diagnosis, prevention, and therapy of bioterrorism threats, including smallpox. The most important and directly relevant of these are summarized below:

- *Approval of Biological Products When Human Efficacy Studies Are Not Ethical or Feasible* (21 CFR 601 Subpart H, as well as 21 CFR 314 Subpart I for New Drugs). This rule, known as "the Animal Rule," was designed to permit approval of drugs and biologics intended to reduce or prevent serious or life-threatening conditions caused by exposure to biological, chemical, radiological, or nuclear substances when human efficacy studies are not ethical and field trials are not feasible (FDA, 2002a).
- *FDA Guidance for Industry—Smallpox (Variola) Infection: Developing Drugs for Treatment or Prevention* (November 2007). This guidance (FDA, 2007a) outlines the unique challenges of developing safe and effective antiviral agents for the treatment and/or prophylaxis of smallpox. These challenges include the exceptionally narrow host range of variola virus, the lack of a previously recognized effective therapeutic agent, and the lack of human diseases that can be considered closely analogous to smallpox.
- The *Project BioShield Act of 2004* (Public Law 108–276). This act establishes a comprehensive Emergency Use Authorization (EUA) program that enables the emergency use of medical products against biological, chemical, radiological, and nuclear attacks, real or potential, for both civilian and military personnel. Under this program, the FDA Commissioner can approve the emergency use of drugs, vaccines, medical devices, and diagnostics not previously approved for a particular purpose (FDA, 2007b).
- The *Public Readiness and Emergency Preparedness Act of 2005* (Public Law 109–148) provides immunity from liability claims aris-

ing from the administration and use of countermeasures covered under EUA.

- The *Public Health Security and Bioterrorism Preparedness and Response Act of 2002* (the "Bioterrorism Act," June 12, 2002) (FDA, 2002b). This act states that the "prompt approval of safe and effective new drugs and other therapies is critical to the improvement of the public health."

The European Union—primarily through the European Medicines Agency (EMEA), which is responsible for the scientific evaluation of applications for European marketing authorization (licensure) of medicinal products in the European Community—has also focused on the threat of bioterrorism in accordance with Article 57(q) of Regulation (EC) No. 726/2004. This article states that the EMEA shall, with a view to protection of the public health, compile "scientific information concerning pathogenic agents which might be used in biological warfare, including the existence of vaccines and other medicinal products available to prevent, or to treat, the effects of such agents." EMEA produced a guidance document in 2002 on the use of available medicinal products for the treatment and prophylaxis of biological agents that might be used as weapons of bioterrorism (European Agency for the Evaluation of Medicinal Products, 2002a). *Note for Guidance on the Development of Vaccinia-Based Vaccines Against Smallpox* applies to the development and manufacture of second-generation vaccinia vaccines produced in embryonated eggs or tissue culture (European Agency for the Evaluation of Medicinal Products, 2002b).

Finally, in 2003 WHO updated its *Recommendations* [formerly known as Requirements] *for the Production and Quality Control of Smallpox Vaccine*, which had last been revised in 1965 (WHO, 2004). The document acknowledges that global resumption of the production of smallpox vaccine would benefit from modern approaches to production and control, and that present-day regulatory expectations should be met in the licensing process. In addition, the document encourages the development of contemporary international reference materials as guidance for determining the potency of new vaccines and their immunogenicity in vaccinated individuals.

STUDY CHARGE AND APPROACH

In anticipation of the WHA meeting in 2010, CDC requested that the IOM convene a committee to conduct a study on the continued use of live variola virus stocks for research and public health purposes. The charge to the committee is presented in Box 1-3.

To address this charge, the IOM convened a committee of experts from both the United States and abroad. Experts in the field of orthopoxvirology

BOX 1-3
Charge to the Committee

An *ad hoc* committee of the Institute of Medicine shall conduct a study on the continued use of live variola virus stocks for research and public health purposes. In follow-on to the IOM's 1999 report, *Assessment of Future Scientific Needs for Live Variola Virus,* an IOM committee will perform a comprehensive evaluation of the research and development work recommended in that report and completed to date, and consider what unmet needs still exist that require the use of live variola virus. The conclusions and recommendations will inform policy discussions in the United States and within the world community regarding the continued need to retain the official stocks of live variola virus for research purposes, and would provide a major review of completed, ongoing and planned research activities that should be undertaken.

The committee shall specifically consider and offer recommendations pertinent to the utility of live variola virus in addressing potential unmet requirements including:

- Advanced development through licensure and post-licensure of antivirals for use in treatment of variola virus infections.
- Advanced development through licensure and post-licensure of new, safe and effective vaccine(s).
- Development through licensure and post-licensure of less-reactogenic vaccines.
- Development of approved protein-based diagnostics which can be used in field situations or diagnostics which have sources of error distinct from those of nucleic acid-based diagnostics.
- Improved pathogenesis data to drive therapeutic discovery.

were consulted, as well as those with expertise in vaccine, antiviral, and diagnostic development; public health; biosecurity; federal government regulation; and bioethics. The committee held two open workshops to gather information from experts and researchers in the salient fields. A comprehensive search of the scientific literature published on variola and other poxviruses was undertaken, and key literature was assessed. The committee also made inquiries to WHO, CDC, and VECTOR regarding research undertaken outside of the United States that might not be readily accessible in the scientific literature.

It is important to note that the committee was charged with assessing scientific needs that require live variola virus. In evaluating unmet needs, the committee recognized the risks of such research and the critical importance of providing independent oversight and essential resources, including BSL-4 facilities when research with live variola virus is undertaken. Commenting on retention or destruction of the live variola virus stocks was not within its scope.

ORGANIZATION OF THE REPORT

The first four chapters of this report provide context for the question of the scientific needs for live variola virus. Following this introductory chapter, Chapter 2 presents an overview of smallpox and its surveillance and control. Chapter 3 examines variola virus in the context of poxvirology and variola's similarities with and differences from other orthopoxviruses. Chapter 4 reviews the state of the art with regard to animal models of the pathogenesis and immunobiology of variola and other poxviruses.

Chapters 5 through 9 review variola-related research completed since the 1999 IOM report was issued, with emphasis on the role of the live virus in advancing scientific breakthroughs. These chapters also address any unmet or future needs in applications of the research, in terms of both medical countermeasures and any additional knowledge that could potentially be gleaned from studying live variola virus. Research in the following areas is examined in turn: genomic analysis (Chapter 5), development of therapeutics (Chapter 6), development of vaccines (Chapter 7), detection of variola and diagnosis of smallpox (Chapter 8), and scientific discovery (Chapter 9). The final chapter presents the committee's conclusions and recommendations.

REFERENCES

Damon, I. K. 2006. Poxviruses. In *Fields' virology*, 5th ed., edited by B. N. Fields, D. M. Knipe, P. M. Howley, and D. E. Griffin. Philadelphia, PA: Lippincott Williams & Wilkins. Pp. 2947–2976.

Dubois, M. E., and M. Slifka. 2008. Retrospective analysis of monkeypox infection. *Emerging Infectious Diseases* 14(4):592–599.

European Agency for the Evaluation of Medicinal Products. 2002a. *EMEA/CPMP guidance document on use of medicinal products for treatment and prophylaxis of biological agents that might be used as weapons of bioterrorism*. CPMP/4048/01. http://www.emea.europa.eu/pdfs/human/bioterror/404801.pdf (accessed January 16, 2009).

European Agency for the Evaluation of Medicinal Products. 2002b. *Note for guidance on the development of vaccinia-based vaccines against smallpox*. http://www.emea.europa.eu/pdfs/human/vwp/110002en.pdf (accessed January 26, 2009).

FDA (Food and Drug Administration). 2002a. New drug and biological drug products; evidence needed to demonstrate effectiveness of new drugs when human efficacy studies are not ethical or feasible. *Federal Register* 67(105). http://www.fda.gov/cber/rules/humeffic.htm (accessed January 13, 2009).

FDA. 2002b. *The Bioterrorism Act of 2002*. http://www.fda.gov/oc/bioterrorism/bioact.html (accessed January 13, 2009).

FDA. 2007a. *Guidance for industry smallpox (variola) infection: Developing drugs for treatment or prevention.* http://www.fda.gov/cder/guidance/7423dft.pdf (accessed January 15, 2009).

FDA. 2007b. *Guidance: Emergency use authorization of medical products.* http://www.fda.gov/oc/guidance/emergencyuse.html (accessed January 18, 2009).

Fenner, F., D. A. Henderson, I. Arita, J. Jezek, and L. D. Ladnyi. 1998. *Smallpox and its eradication.* Geneva, Switzerland: WHO.

IOM (Institute of Medicine). 1999. *Assessment of future scientific needs for live variola virus.* Washington, DC: National Academy Press.

Jenner, E. 1798. *An inquiry into the causes and effects of the variolae vaccinae* (The Classics of Medicine Library). London: Law Murray & Highly.

McCurdy, L. H., B. D. Larkin, J. E. Martin, and B. S. Graham. 2004. Modified vaccinia ankara: Potential as an alternative smallpox vaccine. *Clinical Infectious Diseases* 38(12):1749–1753.

Morens, D. M., G. K. Folkers, and A. S. Fauci. 2008. Emerging infections: A perpetual challenge. *Lancet Infectious Diseases* 8(11):710–719.

Reed, K. D., J. W. Melski, M. B. Graham, R. L. Regnery, M. J. Sotir, M. V. Wegner, J. J. Kazmierczak, E. J. Stratman, Y. Li, J. A. Fairley, G. R. Swain, V. A. Olson, E. K. Sargent, S. C. Kehl, M. A. Frace, R. Kline, S. L. Foldy, J. P. Davis, and I. K. Damon. 2004. The detection of monkeypox in humans in the Western Hemisphere. *New England Journal of Medicine* 350(4):342–350.

Rimoin, A. W., N. Kisalu, B. Kebela-Ilunga, T. Mukaba, L. L. Wright, P. Formenty, N. D. Wolfe, R. L. Shongo, F. Tshioko, E. Okitolonda, J. J. Muyembe, R. W. Ryder, and H. Meyer. 2007. Endemic human monkeypox, Democratic Republic of Congo, 2001–2004. *Emerging Infectious Diseases* 13(6):934–937.

Sale, T. A., J. W. Melski, and J. Strafman. 2006. Monkeypox: An epidemiologic and clinical comparison of African and US disease. *Journal of the American Academy of Dermatology* 55(3):478–481.

Smallpox preservation advisable. 1999. *Nature* 398(6730):733.

Tucker, J. B. 2001. *Scourge: The once and future threat of smallpox.* New York: Atlantic Monthly Press.

WHO (World Health Organization). 1994. *Report of the meeting of the Ad Hoc Committee on Orthopox Virus Infections.* Geneva, Switzerland: WHO.

WHO. 1999. *Report of the meeting of the Ad Hoc Committee on Orthopoxvirus Infections.* January 14–15. Geneva, Switzerland: WHO. http://www.who.int/csr/resources/publications/viral/whocdscsr991.pdf (accessed April 14, 2009).

WHO. 2004. *Recommendations for the production and quality control of smallpox vaccine, revised 2003.* http://www.who.int/biologicals/publications/trs/areas/vaccines/smallpox/Annex%201%20(27-64)TRS926Smallpox2003.pdf (accessed January 26, 2009).

WHO. 2008. *WHO Advisory Committee on Variola Virus Research Report of the Tenth Meeting.* November 19–20. Geneva, Switzerland: WHO. http://www.who.int/csr/resources/publications/WHO_HSE_EPR_2008_9/en/index.html (accessed March 23, 2009).

2

Overview of Smallpox and Its Surveillance and Control

Smallpox, the disease caused by the variola virus, is characterized by fever; headache; back pain; vomiting; and, most distinctly, a papular, and later vesicular, rash. Smallpox has a lengthy incubation period that averages 12–14 days, during which time the infected person is non-contagious. Within 2–3 days of the sudden onset of fever and other symptoms, skin lesions begin to appear on the face, hands, arms, and legs, and eventually the trunk. Lesions erupt first on mucosal surfaces, including the mouth and nasal cavities, where they ulcerate and shed the virus in respiratory secretions (see Figure 2-1). Smallpox is most contagious during the febrile period and early stages of the rash, but remains transmissible until the resulting scabs have fallen off (Breman and Henderson, 2002).

Smallpox was originally considered a single disease. However, it was subsequently subdivided into two clinical types, caused by closely related variants of the variola virus: "classical" or variola major, and variola minor or alastrim. The former had a higher case fatality rate of around 30 percent, while the latter was less severe, with only about 1 percent of cases resulting in death (Henderson and Fenner, 2001).

EPIDEMIOLOGY

Smallpox is uniquely a human disease, and variola virus has no other known host or reservoir species. Historically, the virus was transmitted primarily through aerosolization of respiratory secretions, as well as by direct contact with skin lesions or exposure to contaminated bedding or clothing. For variola major, transmission occurred mainly to close contacts because

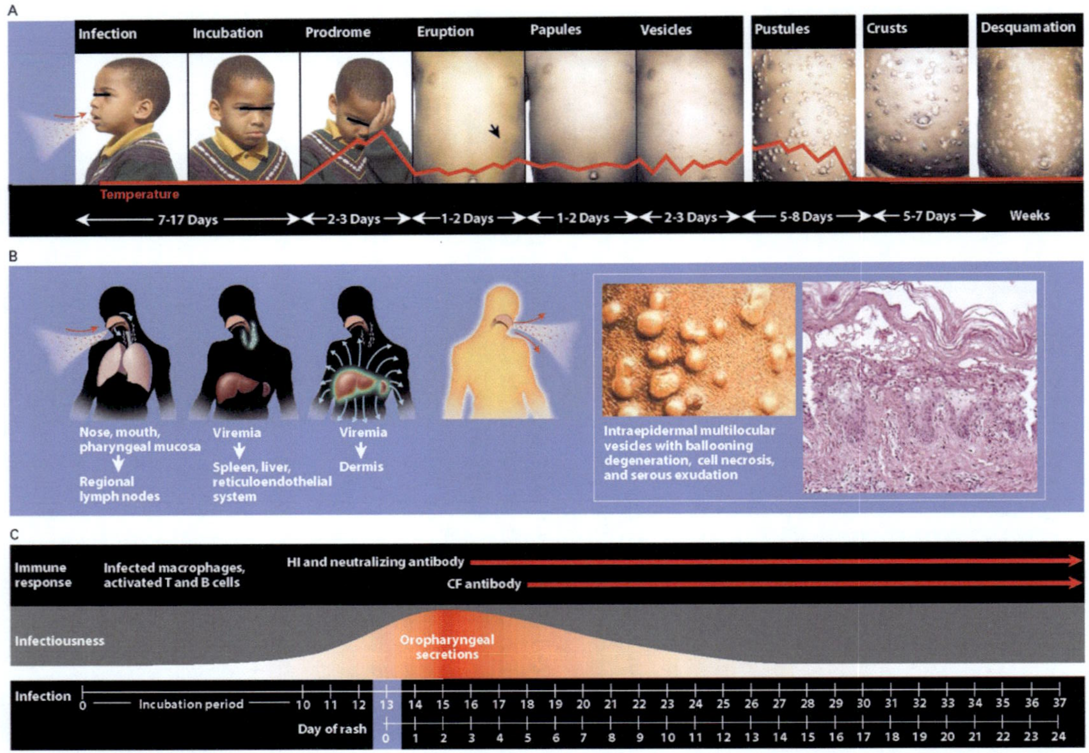

FIGURE 2-1 Clinical manifestations and pathogenesis of smallpox and the immune response. Reprinted with permission from (Breman and Henderson, 2002) and (Strano, 1976). Copyright © 2002 Massachusetts Medical Society. All rights reserved. Other images provided by WHO, NIH, the American Registry of Pathology.

the severity of the disease rendered most victims bed-ridden shortly after the onset of illness. Variola minor, with its milder presentation, could be transmitted much more widely because of patients' mobility and remained endemic in some parts of the world even after variola major had been eliminated (Fenner et al., 1998).

Smallpox epidemics occurred in cycles that varied from annually to every few years. The periodicity depended largely on the number of susceptible individuals in the community, which was heavily influenced by the prevalence of prior infection and by vaccination levels (Fenner et al., 1998). As smallpox vaccination coverage increased, the size and frequency of outbreaks decreased (Fenner et al., 1998).

Smallpox was endemic in almost all parts of the world until the mid-twentieth century. Vaccination campaigns had eliminated the disease from nearly all of Europe, Australia, and New Zealand by the early 1950s and from the American continents a decade later. The last case of smallpox in the United States occurred in 1949. Global eradication efforts accelerated in the mid-1960s, and areas of endemicity rapidly diminished in Asia and Africa. As noted in Chapter 1, the last known naturally transmitted case of smallpox occurred in 1977 in Somalia, while the last known case of the disease was due to a laboratory-associated accident in England the following year. WHO declared smallpox eradicated in May 1980. This achievement has not yet been repeated with any other human pathogen. Table 2-1 summarizes the timeline for smallpox eradication.

SURVEILLANCE AND CONTROL

The 2001 anthrax attacks in the United States reminded the world that a biological agent could be used as a weapon of terror and made the research agenda for high-consequence pathogens such as variola a national priority (Lane et al., 2001). Even though naturally occurring smallpox has been eradicated (Henderson, 1987), the risk of smallpox resulting from a deliberate or accidental release of the agent remains (Mahalingam et al., 2004).

Because of its characteristic rash, surveillance for smallpox was straightforward when natural disease was present in the world. Today, by contrast, physicians lack familiarity with smallpox and may be unable to diagnose it (Breman and Henderson, 2002; Woods et al., 2004). WHO considers a single verified case of smallpox to be a public health emergency of international concern, and under the 2005 revisions of the International Health Regulations, reporting of such a case to WHO is obligatory. A diagnosis of smallpox must be confirmed by laboratory testing. Whereas transmission was historically limited primarily to close contacts, most people now alive have no natural or vaccine-induced immunity to the disease, and society is

TABLE 2-1 Timeline for Smallpox Eradication

Date	Location	Event
430 BC		Survivors of smallpox called upon to care for the afflicted (as survivors were immune)
Unknown		Variolation, or inoculation, practiced in Africa, India, and China
1721	Europe and North America	Variolation method introduced
1744	Japan	Variolation method introduced
1798	England	Edward Jenner first to discover a vaccine using cowpox
1909	Guinea	First time an experimental dried vaccine was used
1949	Michigan State Laboratories	Freeze-drying invented
1949	United States	Last case of smallpox
1950s	Western Hemisphere	Eradication program started in western hemisphere by Pan American Sanitary Organization
1954	Lister Institute in England	Freeze-dried vaccine produced for commercial use
1958		USSR suggests a global eradication program to WHA
1966		WHA decides to intensify the eradication program
1967		Intensified plan for eradication is launched by WHO
1977	Somalia	Last naturally occurring case in the world
1978	United Kingdom	Last two cases in the world, laboratory acquired
1979		Global eradication certified by a group of scientists
1980	Global	Eradication and previous certification endorsed by WHA

highly mobile; therefore, transmission dynamics today may be considerably different from those seen in the past.

One key to implementing effective disease control strategies for a pathogen such as variola is prompt and accurate detection, either directly by identifying the biological agent or indirectly by methods that demonstrate the host's response to the suspected pathogen (Fraser et al., 2004). Since 1999, technological advances have yielded laboratory methods that permit the analysis of clinical specimens for orthopoxvirus nucleic acid (Loparev et al., 2001; Nitsche et al., 2004; Olson et al., 2004; Wenli et al., 2004; Aitichou et al., 2005; Shchelkunov et al., 2005; Fitzgibbon et al., 2006; Li et al., 2007; Sulaiman et al., 2008) or orthopoxvirus-specific proteins or antibodies (Karem et al., 2005; Huelseweh et al., 2006; Davies et al., 2007). CDC has distributed validated variola clinical diagnostics through the Laboratory Response Network (LRN), and assays for environmental detection exist (CDC, 2008).

In response to the detection of variola, three options exist for controlling any resulting outbreak of disease: isolation and quarantine, vaccination, and administration of antiviral drugs. CDC has specific procedures in place for containment of the disease should it be diagnosed, including use of isolation and quarantine, identification and vaccination of close contacts, and vaccination of those not directly exposed. Similar protocols exist elsewhere in the world.

The last decade has seen considerable efforts to develop next-generation smallpox vaccines, and progress has been made in the development and licensure of live attenuated vaccinia-based vaccines utilizing modern production techniques (Monath et al., 2004; Vollmar et al., 2006; Wiser et al., 2007; Artenstein, 2008; Greenberg and Kennedy, 2008). In addition, contemporary experience has been acquired with vaccinating large populations of individuals, including military personnel (CIDRAP, 2008) and volunteer first responders and laboratory workers (Casey et al., 2005). This experience has yielded new data on the safety profile and adverse effects associated with vaccination in a largely immunologically naïve population (Fulginiti et al., 2003; Grabenstein and Winkenwerder, 2003; Halsell et al., 2003; Talbot et al., 2003; Greenberg et al., 2004; Wollenberg and Engler, 2004; Malone, 2007; Kroger et al., 2008; Reif et al., 2008), as well as on the nature of the host's response (Hammarlund et al., 2003a,b; Kennedy et al., 2004; Kim et al., 2006, 2007; Kan et al., 2007; Gassmann et al., 2008; Grosenbach et al., 2008). Progress has also been made in the development of drugs for treatment and postexposure prophylaxis of smallpox (Yang et al., 2005; Sliva and Schnierle, 2007; Bolken and Hruby, 2008; Nalca et al., 2008; Tse-Dinh, 2008; Painter et al., 2008).

Despite the research that has been accomplished since 1999, capability gaps for smallpox control remain. These include the development and licensure of rapid field diagnostics that are specific for variola or for antibodies induced by variola infection, further assessment and licensure of antivirals for the treatment of smallpox, and a licensed smallpox vaccine with a more favorable safety profile.

REFERENCES

Aitichou, M., S. Javorschi, and M. S. Ibrahim. 2005. Two-color multiplex assay for the identification of orthopox viruses with real-time LUX-PCR. *Molecular & Cellular Probes* 19(5):323–328.

Artenstein, A. W. 2008. New generation smallpox vaccines: A review of preclinical and clinical data. *Reviews in Medical Virology* 18(4):217–231.

Bolken, T. C., and D. E. Hruby. 2008. Discovery and development of antiviral drugs for biodefense: Experience of a small biotechnology company. *Antiviral Research* 77(1):1–5.

Breman, J. G., and D. A. Henderson. 2002. Diagnosis and management of smallpox. *New England Journal of Medicine* 346(17):1300–1308.

Casey, C. G., J. K. Iskander, M. H. Roper, E. E. Mast, X. J. Wen, T. J. Torok, L. E. Chapman, D. L. Swerdlow, J. Morgan, J. D. Heffelfinger, C. Vitek, S. E. Reef, L. M. Hasbrouck, I. Damon, L. Neff, C. Vellozzi, M. McCauley, R. A. Strikas, and G. Mootrey. 2005. Adverse events associated with smallpox vaccination in the United States, January–October 2003. *The Journal of the American Medical Association* 294(21).

CDC (Centers for Disease Control and Prevention). 2008. *Acute generalized vesicular or pustular rash illness testing protocol in the United States.* http://emergency.cdc.gov/agent/smallpox/diagnosis/pdf/poxalgorithm11-14-07.pdf (accessed February 13, 2008).

CIDRAP (Center for Infectious Disease Research and Policy). 2008. *US military switching to new smallpox vaccine.* http://www.cidrap.umn.edu/cidrap/content/bt/smallpox/news/feb0808smallpox.html (accessed April 15, 2009).

Davies, D. H., D. M. Molina, J. Wrammert, J. Miller, S. Hirst, Y. Mu, J. Pablo, B. Unal, R. Nakajima-Sasaki, X. Liang, S. Crotty, K. L. Karem, I. K. Damon, R. Ahmed, L. Villarreal, and P. L. Felgner. 2007. Proteome-wide analysis of the serological response to vaccinia and smallpox. *Proteomics* 7(10):1678–1686.

Fenner, F., D. A. Henderson, I. Arita, J. Jezek, and L. D. Ladnyi. 1998. *Smallpox and its eradication.* Geneva, Switzerland: WHO.

Fitzgibbon, J. E., and J. L. Sagripanti. 2006. Simultaneous identification of orthopoxviruses and alphaviruses by oligonucleotide macroarray with special emphasis on detection of variola and Venezuelan equine encephalitis viruses. *Journal of Virological Methods* 131(2):160–167.

Fraser, C., S. Riley, R. M. Anderson, and N. M. Ferguson. 2004. Factors that make an infectious disease outbreak controllable. *Proceedings of the National Academy of Sciences of the United States of America* 101(16):6146–6151.

Fulginiti, V. A., A. Papier, J. M. Lane, J. M. Neff, and D. A. Henderson. 2003. Smallpox vaccination: A review, part II. Adverse events. *Clinical Infectious Diseases* 37(2):251–271.

Gassmann, R., O. B. Engler, R. Steffen, M. Alex, C. P. Czerny, and M. Mutsch. 2008. Clinical and immune response to undiluted and diluted smallpox vaccine. *Swiss Medical Weekly* 138:392–397.

Grabenstein, J. D., and W. Winkenwerder, Jr. 2003. US military smallpox vaccination program experience. *The Journal of the American Medical Association* 289(24):3278–3282.

Greenberg, R. N., and J. S. Kennedy. 2008. ACAM2000: A newly licensed cell culture-based live vaccinia smallpox vaccine. *Expert Opinion on Investigational Drugs* 17(4):555–564.

Greenberg, R. N., R. H. Schosser, E. A. Plummer, S. E. Roberts, M. A. Caldwell, D. L. Hargis, D. W. Rudy, M. E. Evans, and R. J. Hopkins. 2004. Urticaria, exanthems, and other benign dermatologic reactions to smallpox vaccination in adults. *Clinical Infectious Diseases* 38(7):958.

Grosenbach, D. W., R. Jordan, D. S. King, A. Berhanu, T. K. Warren, D. L. Kirkwood-Watts, S. Tyavanagimatt, Y. Tan, R. L. Wilson, K. F. Jones, and D. E. Hruby. 2008. Immune responses to the smallpox vaccine given in combination with ST-246, a small-molecule inhibitor of poxvirus dissemination. *Vaccine* 26(7):933–846.

Halsell, J. S., J. R. Riddle, J. E. Atwood, P. Gardner, R. Shope, G. A. Poland, G. C. Gray, S. Ostroff, R. E. Eckart, D. R. Hospenthal, R. L. Gibson, J. D. Grabenstein, M. K. Arness, D. N. Tornberg, and Team Department of Defense Smallpox Vaccination Clinical Evaluation. 2003. Myopericarditis following smallpox vaccination among vaccinia-naive US military personnel. *The Journal of the American Medical Association* 289(24):3283–3289.

Hammarlund, E., M. W. Lewis, S. G. Hansen, L. I. Strelow, J. A. Nelson, G. J. Sexton, J. M. Hanifin, and M. K. Slifka. 2003a. Duration of antiviral immunity after smallpox vaccination. *Nature Medicine* 9(9):1131–1137.

Hammarlund, E. K., M. W. Lewis, J. M. Hanifin, and M. K. Slifka. 2003b. Is there still immunity to smallpox? *Abstracts of the General Meeting of the American Society for Microbiology* 103(086).

Henderson, D. A. 1987. Principles and lessons from the smallpox eradication programme. *Bulletin of the World Health Organization* 65(4):535–546.

Henderson, D. A., and F. Fenner. 2001. Recent events and observations pertaining to smallpox virus destruction in 2002. *Clinical Infectious Diseases* 33(7):1057–1059.

Huelseweh, B., R. Ehricht, and H.-J. Marschall. 2006. A simple and rapid protein array based method for the simultaneous detection of biowarfare agents. *Proteomics* 6(10):2972–2981.

Kan, V. L., J. Manischewitz, L. R. King, and H. Golding. 2007. Durable neutralizing antibodies after remote smallpox vaccination among adults with and without HIV infection. *AIDS (London, England)* 21(4):521–524.

Karem, K. L., M. Reynolds, Z. Braden, G. Lou, N. Bernard, J. Patton, and I. K. Damon. 2005. Characterization of acute-phase humoral immunity to monkeypox: Use of immunoglobulin M enzyme-linked immunosorbent assay for detection of monkeypox infection during the 2003 North American outbreak. *Clinical and Diagnostic Laboratory Immunology* 12(7):867–872.

Kennedy, J. S., S. E. Frey, L. Yan, A. L. Rothman, J. Cruz, F. K. Newman, L. Orphin, R. B. Belshe, and F. A. Ennis. 2004. Induction of human T cell-mediated immune responses after primary and secondary smallpox vaccination. *Journal of Infectious Diseases* 190(7):1286–1294.

Kim, S. H., S. G. Yeo, J. H. Cho, H. B. Kim, N. J. Kim, M. D. Oh, K. W. Choe, Y. Jee, and H. Cho. 2006. Cell-mediated immune responses to smallpox vaccination. *Clinical and Vaccine Immunology* 13(10):1172–1174.

Kim, S. H., S. J. Choi, W. B. Park, H. B. Kim, N. J. Kim, M. d Oh, and K. W. Choe. 2007. Detailed kinetics of immune responses to a new cell culture-derived smallpox vaccine in vaccinia-naive adults. *Vaccine* 25(33):6287–6291.

Kroger, A., C. Vellozzi, M. Deming, C. G. Casey, X. Wen, and S. A. Norton. 2008. Dermatological lesions near the smallpox vaccination site after scab detachment. *Clinical Infectious Diseases* 46(Suppl. 3):S227–S233.

Lane, H. C., J. L. Montagne, and A. S. Fauci. 2001. Bioterrorism: A clear and present danger. *Nature Medicine* 7(12):1271–1273.

Li, Y., S. L. Ropp, H. Zhao, I. K. Damon, and J. J. Esposito. 2007. Orthopoxvirus pangenomic DNA assay. *Journal of Virological Methods* 141(2):154–165.

Loparev, V. N., R. F. Massung, J. J. Esposito, and H. Meyer. 2001. Detection and differentiation of Old World orthopoxviruses: Restriction fragment length polymorphism of the crmB gene region. *Journal of Clinical Microbiology* 39(1):94–100.

Mahalingam, S., I. K. Damon, and B. A. Lidbury. 2004. 25 years since the eradication of smallpox: Why poxvirus research is still relevant. *Trends in Immunology* 25(12):636–639.

Malone, J. D. 2007. Pre-event smallpox vaccination for healthcare workers revisited—the need for a carefully screened multidisciplinary cadre. *International Journal of Infectious Diseases* 11(2):93–97.

Monath, T. P., J. R. Caldwell, W. Mundt, J. Fusco, C. S. Johnson, M. Buller, J. Liu, B. Gardner, G. Downing, P. S. Blum, T. Kemp, R. Nichols, and R. Weltzin. 2004. ACAM2000 clonal Vero cell culture vaccinia virus (New York City Board of Health strain)—a second-generation smallpox vaccine for biological defense. *International Journal of Infectious Diseases* 8(Suppl. 2):S31–S44.

Nalca, A., J. M. Hatkin, N. L. Garza, D. K. Nichols, S. W. Norris, D. E. Hruby, and R. Jordan. 2008. Evaluation of orally delivered ST-246 as postexposure prophylactic and antiviral therapeutic in an aerosolized rabbitpox rabbit model. *Antiviral Research* 79(2):121–127.

Nitsche, A., H. Ellerbrok, and G. Pauli. 2004. Detection of orthopoxvirus DNA by real-time PCR and identification of variola virus DNA by melting analysis. *Journal of Clinical Microbiology* 42(3):1207–1213.

Olson, V. A., T. Laue, M. T. Laker, I. V. Babkin, C. Drosten, S. N. Shchelkunov, M. Niedrig, I. K. Damon, and H. Meyer. 2004. Real-time PCR system for detection of orthopoxviruses and simultaneous identification of smallpox virus. *Journal of Clinical Microbiology* 42(5):1940–1946.

Painter, G. R., L. C. Trost, B. M. Lampert, M. R. Almond, R. M. Buller, E. Kern, G. P. Painter, A. T. Robertson, and R. O'Mahony. 2008. CMX001. *Drugs of the Future* 33(8):655.

Reif, D. M., B. A. McKinney, A. A. Motsinger, S. J. Chanock, K. M. Edwards, M. T. Rock, J. H. Moore, and J. E. Crowe, Jr. 2008. Genetic basis for adverse events after smallpox vaccination. *The Journal of Infectious Diseases* 198(1):16–22.

Shchelkunov, S. N., E. V. Gavrilova, and I. V. Babkin. 2005. Multiplex PCR detection and species differentiation of orthopoxviruses pathogenic to humans. *Molecular & Cellular Probes* 19(1):1–8.

Sliva, K., and B. Schnierle. 2007. From actually toxic to highly specific: Novel drugs against poxviruses. *Virology Journal* 4(8).

Strano, A.J. 1976. *Smallpox.* In: pathology of Tropical and Extraordinary Diseases. American Registry of Pathology: Washington, DC.

Sulaiman, I. M., S. A. Sammons, and R. M. Wohlhueter. 2008. Smallpox virus resequencing GeneChips can also rapidly ascertain species status for some zoonotic non-variola orthopoxviruses. *Journal of Clinical Microbiology* 46(4):1507–1509.

Talbot, T. R., H. K. Bredenberg, M. Smith, B. J. Lafleur, A. Boyd, and K. M. Edwards. 2003. Focal and generalized folliculitis following smallpox vaccination among vaccinia-naive recipients. *The Journal of the American Medical Association* 289(24):3290–3294.

Tse-Dinh, Y. C. 2008. An update on the development of drugs against smallpox. *Current Opinion in Investigational Drugs* 9(8):865–870.

Vollmar, J., N. Arndtz, K. M. Eckl, T. Thomsen, B. Petzold, L. Mateo, B. Schlereth, A. Handley, L. King, V. Hulsemann, M. Tzatzaris, K. Merkl, N. Wulff, and P. Chaplin. 2006. Safety and immunogenicity of IMVAMUNE, a promising candidate as a third generation smallpox vaccine. *Vaccine* 24(12):2065–2070.

Wenli, M., W. Yan, W. Hongmin, and Z. Wenling. 2004. An oligonucleotide microarray for the detection of vaccinia virus. *British Journal of Biomedical Science* 61(3):142–145.

Wiser, I., R. D. Balicer, and D. Cohen. 2007. An update on smallpox vaccine candidates and their role in bioterrorism related vaccination strategies. *Vaccine* 25(6):976–984.

Wollenberg, A., and R. Engler. 2004. Smallpox, vaccination and adverse reactions to smallpox vaccine. *Current Opinion in Allergy & Clinical Immunology* 4(4):271–275.

Woods, R., T. McCarthy, M. A. Barry, and B. Mahon. 2004. Diagnosing smallpox: Would you know it if you saw it? *Biosecurity & Bioterrorism* 2(3):157–163.

Yang, G., D. C. Pevear, M. H. Davies, M. S. Collett, T. Bailey, S. Rippen, L. Barone, C. Burns, G. Rhodes, S. Tohan, J. W. Huggins, R. O. Baker, R. L. M. Buller, E. Touchette, K. Waller, J. Schriewer, J. Neyts, E. DeClercq, K. Jones, D. Hruby, and R. Jordan. 2005. An orally bioavailable antipoxvirus compound (ST-246) inhibits extracellular virus formation and protects mice from lethal orthopoxvirus challenge. *Journal of Virology* 79(20):13139–13149.

3

Comparative Poxvirology

A comprehensive discussion of variola virus must include comparison with other orthopoxviruses, as the similarities and differences between variola and members of its family help elucidate how variola causes disease and how it is modulated by subsequent host responses. In particular, variola virus's unique adaptation to a single host (while other orthopoxviruses readily infect multiple mammalian species) and its ability to induce unusually severe disease hint at a complex relationship between host and pathogen that may not be as easily explored using other orthopoxviruses. Variola virus also is not a single virus whose strains are identical, and the implications of this diversity could yield further information not just about the host, but also about mechanisms of antiviral therapeutic action, vaccine efficacy, and rapid diagnostic capabilities. Taking advantage of insights from comparative virology is particularly important because experiments with live variola virus must focus on critical questions. Knowledge of the related poxviruses can inform the design and refinement of experiments for which live variola virus is necessary. In the last decade, technological advances and the development of molecular techniques have made it possible to gain a deeper understanding of the general mechanisms involved in poxvirus replication, the host response, and the ways in which these pathogens have adapted to their hosts that is pertinent as background for considering the scientific needs for variola virus.

POXVIRUS TAXONOMY

Viruses with shared characteristics are grouped into taxonomic categories, including those of the *Poxviridae* family. Variola and the other members of *Poxviridae* are among the largest and most complex known viruses. Their genome is a single linear, double-stranded DNA molecule between 130 and 360 kilobase pairs (kbp) in size, and encodes on average approximately 150 proteins. The family *Poxviridae* is subdivided into two subfamilies based on the restriction of their host range to vertebrates (*Chordopoxvirinae*) or invertebrates (*Entomopoxvirinae*), and these subfamilies are further subdivided into genera of viruses that are genetically related and share aspects of nucleotide composition, host range, and morphology. *Chordopoxvirinae* consists of eight genera: *Orthopoxvirus*, *Parapoxvirus*, *Avipoxvirus*, *Capripoxvirus*, *Leporipoxvirus*, *Suipoxvirus*, *Molluscipoxvirus*, and *Yatapoxvirus* (Moss, 2007). The *Orthopoxvirus* genus includes many of the known poxviruses that naturally infect mammals, including vaccinia, the prototypical poxvirus, as well as two of the three poxviruses that have most commonly infected humans, variola and monkeypox virus; the third, molluscum contagiosum virus, is a member of *Molluscipoxvirus*. Other members of *Chordopoxvirinae*, including cowpox virus (genus *Orthopoxvirus*) and orf virus (genus *Parapoxvirus*, most commonly found in sheep and goats) are less common disease agents of humans. Table 3-1 lists the poxviruses that affect humans, along with their reservoir hosts, other infected hosts, and geographic distribution.

POXVIRUS STRUCTURE

Poxviruses were first visualized by electron microscopy (EM) in 1938 (Biel and Gelderblom, 1999). The large virus particles (approximately 240 nm × 300 nm for orthopoxviruses) appear brick-shaped under standard EM, with internal structures resembling a dumbbell-shaped core and two lateral bodies (see Figure 3-1). From the late 1940s through the end of the smallpox eradication era, EM was used to diagnose smallpox and to differentiate between variola and varicella zoster virus (VZV), which causes chickenpox (Biel and Gelderblom, 1999). The distinctive morphology observed by EM can still be a first step in diagnosis of poxviruses. During the outbreak of monkeypox in the United States in 2003, the first realization that the etiologic agent was an orthopoxvirus occurred when brick-shaped virions were visualized in a clinical specimen by EM (Reed et al., 2004). However, it is important to recognize that the orthopoxviruses that infect humans, including variola, vaccinia, and monkeypox, cannot be differentiated by traditional EM alone because the virion structure is highly conserved among orthopoxviruses.

POXVIRUS GENOMICS

Advances in genomic sequencing and computational molecular biology have provided new insights into the relatedness and evolutionary history

TABLE 3-1 Poxviruses That Infect Humans

Genus	Virus	Reservoir Hosts	Other Infected Hosts	Geographic Distribution
Orthopoxvirus	Cowpox	Bank voles, long-tailed field mice	Humans, cats, cattle, zoo animals	Europe, western Africa
	Monkeypox	Unknown, likely rodents	Humans, monkeys, zoo animals, prairie dogs	Western and central Africa
	Vaccinia	?	Humans, rabbits, cattle, river buffalo	?
	Variola	Humans	None	Eradicated (formerly worldwide)
Parapoxvirus	Bovine papular stomatitis	Cattle (beef)	Humans	Worldwide
	Orf	Sheep, goats	Humans, ruminants	Worldwide
	Pseudocowpox	Cattle (dairy)	Humans	Worldwide
	Sealpox	Seals	Humans	Worldwide
Yatapoxvirus	Tanapox		Humans	Eastern and central Africa
	Yabapox	? Primates	Humans	Western Africa
Molluscipoxvirus	Molluscum contagiosum	Humans	None	Worldwide

of poxviruses. This information helps place variola in its evolutionary context and points to significant genetic differences between variola and other orthopoxviruses. Poxvirus genes are usually nonoverlapping but closely spaced, and are arranged in blocks such that genes in the outer quadrants of the genome are transcribed toward the end of the genome in closest proximity, while genes in the central quadrants are transcribed toward the center of the genome. An analysis of 21 *Poxviridae* complete genome sequences in 2003 revealed a common set of 49 genes and an additional 41 genes shared by the chordopoxviruses (Upton et al., 2003). These families of shared genes encode proteins involved in basic functions such as DNA replication, transcription, and virion assembly, and are located toward the central region of the genome. In contrast, genes that are virus- or host-specific tend to be located toward the genome termini, and encode factors

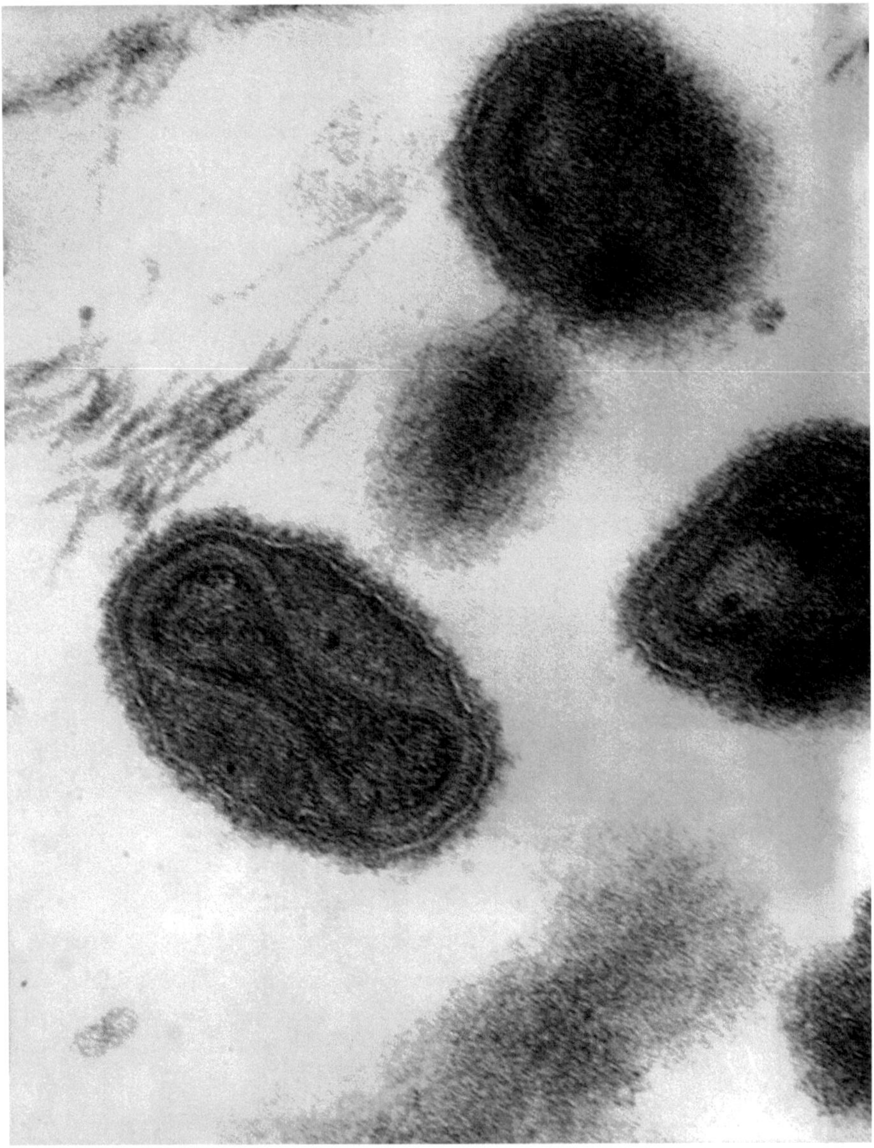

FIGURE 3-1 Electron microscopy of orthopoxvirus structure.

involved in subversion of host defenses and immune responses. These virulence-associated genes are believed to have been acquired more recently by the virus as it adapted to the host species (Lefkowitz et al., 2006). To date, all poxvirus genomes that have been studied have been found to have inverted terminal repeats (ITRs) at both ends of the genome (Garon et al., 1978; see Figure 3-2).

Contemporary taxonomic approaches combine sequence-based phylogenetic and character trait analyses. This is the case for the poxviruses (Lefkowitz et al., 2006). Alignments of concatenated orthologous protein sequences from the poxviruses have led to reconsideration of genus assignments for some members and genus interrelationships. Gene loss, fragmentation, and duplication all appear to have played important roles in poxvirus evolution, with subsequent restriction of virus host range. Complete genome sequences and comparative analysis have suggested the basis for differences in virulence among strains of the same orthopoxvirus species, as illustrated by monkeypox virus (MPXV). In recent years, it has become increasingly well recognized that west African strains of MPXV are less virulent than central African (Congo basin) strains, despite roughly similar degrees of host exposure in these two regions of the continent. The genome sequences of three west African MPXV strains were found to be more closely related to each other (0.01–0.07 percent difference) than to the previously sequenced strain from the Democratic Republic of the Congo (0.55–0.56 percent difference) (Chen et al., 2005). Of note, five putative virulence-associated genes contained significant deletions or fragmenta-

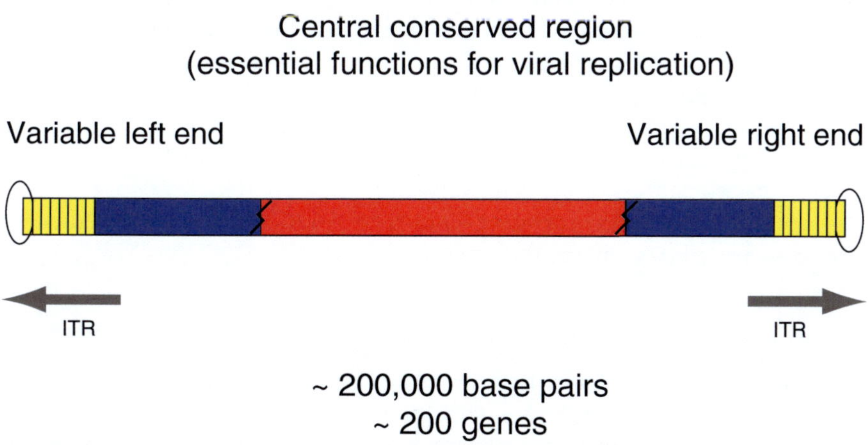

FIGURE 3-2 Internal terminal repeats at both ends of the poxvirus genome.

tion in the three west African strains as compared with the central African strain, including the gene encoding the MPXV ortholog of the vaccinia complement-binding protein (VCP-MPXV). On the basis of this sequence analysis and subsequent assessment of the expressed protein, VCP-MPXV is hypothesized to play a key role in the virulence of west African MPXV strains. More genomic information about variola virus isolates, coupled with clinical data on disease severity in the cases from which the virus was recovered, has the potential to elucidate factors influencing the virulence of variola.

VIRAL LIFE CYCLE

Variola virus, like all orthopoxviruses, replicates solely in the cytoplasm of infected cells (see Figure 3-3). Most of what is known about the orthopoxvirus life cycle has been learned from extensive study of vaccinia virus (Moss, 2007), which is closely related to variola virus. Vaccinia has been the prototypic model for experimental analysis of the orthopox viral life cycle in vitro. Not only is analysis of the vaccinia life cycle relevant for understanding variola, but because smallpox vaccines are made from vaccinia, this information can be used to design safer alternatives.

There are several key points to be kept in mind as the viral life cycle is reviewed. First, progression through this life cycle involves the action of a significant number of virally encoded products, which are largely conserved throughout the *Orthopoxvirus* genus but do exhibit strain-specific variations. Subtle features of the variola proteome are likely to be important in this regard. Second, there is increasing evidence—particularly for the process of transcription—that host proteins participate in the viral life cycle. This participation of host proteins is likely to be a significant determinant of species specificity and may contribute to the narrow host range of variola. Third, the progression of the viral life cycle occurs within the context of the cell, and there is growing evidence that intracellular structures (e.g., cytoskeleton, membranes) are vital for efficient viral replication. How viral proteins interact with these cellular structures is likely to be strain- and species-specific, and this is a relatively understudied area. Fourth, the viral proteins that mediate the progression of the viral life cycle represent the key pool of targets for antiviral therapy.

Viral Entry

There is good evidence that mature virions (MV) form weak attachments with glycosaminoglycans and laminins on the cell surface; various proteins within the virion membrane have been shown to be responsible for these interactions (Chung et al., 1998; Carter et al., 2005; Chiu et al.,

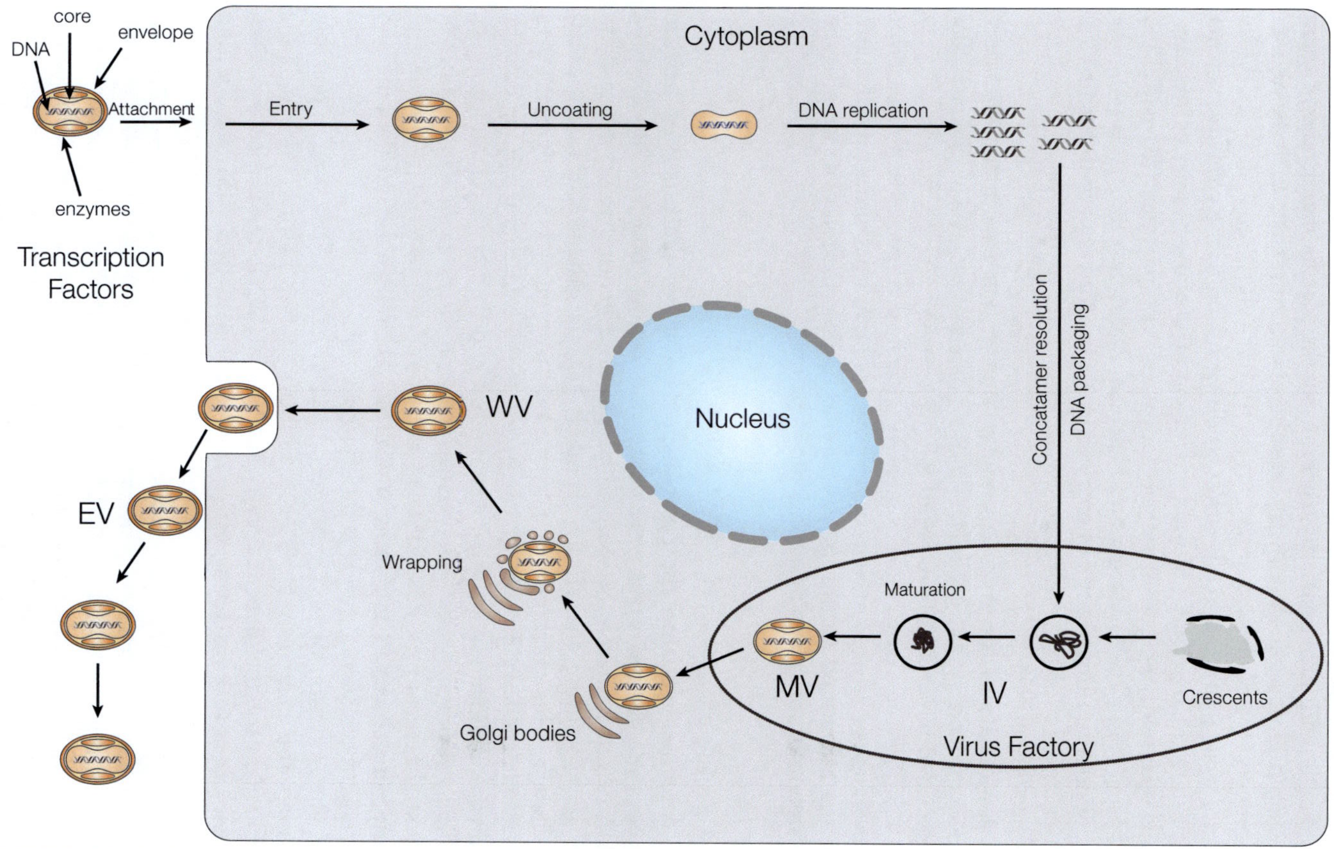

FIGURE 3-3 Orthopoxvirus replication.

2007). It is not yet known whether other viral proteins interact strongly or specifically with proteinaceous receptors on the plasma membrane. If there are such receptors, however, they must be ubiquitous, since there have been no reports of cell lines that cannot support viral binding or entry. Once bound, virions can enter cells either by direct fusion of the viral and plasma membranes or by uptake of intact virus via macropinocytosis, with subsequent release of the core from the endosomal compartment (Moss, 2006; Townsley et al., 2006; Mercer and Helenius, 2008) into the cytoplasm. The predominance of one mode of entry or the other differs with different viral strains and cell types, although the determinants of this variability have not been identified.

A minority of the virions produced during poxvirus infection mature into enveloped or extracellular virions, also known as EV. EV are MV surrounded by an additional lipid bilayer carrying EV-specific surface proteins (Smith and Law, 2004). When EV attach to target cells, this exterior membrane is ruptured in a process known as ligand-dependent dissolution, and the MV found within then enter cells as described above (see also the discussion of morphogenesis and egress below).

Gene Expression

After delivery into the cytoplasm, the core of the poxvirus particle remains intact, and early gene expression begins (Broyles, 2003; Moss, 2007). Approximately 50 percent of the viral genes are expressed from the genome using the virally encoded and encapsidated transcriptional machinery. This machinery includes a multisubunit RNA polymerase, an RNA Pol accessory protein, an early transcription factor, a capping enzyme and cap modification enzyme, termination and transcript release proteins, and poly A polymerase. This phase in the viral life cycle yields a number of potential targets for antiviral drugs that may be conserved in variola and related poxviruses. The mature transcripts are released from the core and undergo translation on host cell polysomes.

The initiation of two subsequent phases of poxvirus gene expression requires the prior onset of DNA replication (see below). Although the mechanism for this dependency is not known, DNA replication has been shown to have a *cis*-acting effect on the encapsidated viral genome that enables intermediate and late transcription to occur. Intermediate gene expression utilizes viral transcription factors that are expressed as early proteins, as well as some host proteins. Late gene expression also uses distinct viral transcription factors, as well as some host factors. Two features that distinguish these postreplicative phases of gene expression are the presence of 5′ polyA heads on the transcripts and imprecise termination of transcription; the latter feature leads to the presence of long overlapping transcripts, which leads in turn to the presence of dsRNA. The availability of distinct classes of viral mRNA at distinct times after infection is also enhanced by the expression of viral decapping enzymes, which accelerate mRNA turnover (Parrish and Moss, 2007; Parrish et al., 2007).

The cytoplasmic replication cycle of poxviruses means that both transcription and translation occur in the cytoplasm and can be both temporally and physically coupled. Indeed, postreplicative transcription and translation occur within distinct areas of the cytoplasm known as viral factories (Condit, 2007; Katsafanas and Moss, 2007). This compartmentalization serves to enhance viral protein expression as well as to diminish cellular protein expression, since the translational machinery is depleted in the cytoplasmic areas outside of the viral factories.

Genome Replication and Maturation

After early gene expression, the core appears to disassemble and the genome is released into the cytoplasm, where it undergoes replication in the viral factories. The proteins that make up the viral replication machinery, all of which are conserved in variola, are expressed early after infection. They include a catalytic DNA polymerase, a dimeric processivity factor (one subunit of which is an enzymatically active uracil DNA glycosylase), an ssDNA binding protein, and a primase/helicase (Moss and De Silva, 2006). A virally encoded protein kinase is also essential for replication; its primary role appears to be to phosphorylate and thus overcome the inhibitory action of a cytoplasmic DNA sensor. Other accessory proteins that may be dispensable in tissue culture but essential in vivo are also encoded by the viral genome. The mechanism by which replication initiates is still in question, and there is as yet no clear answer as to whether replication involves only leading strand synthesis or both leading and lagging strand synthesis. Replication does lead to the synthesis of tail/tail concatemers of the genome, which undergo subsequent resolution to mature monomeric genomes. This resolution is accomplished by a virally encoded Holliday junction resolvase and leads to reformation of the unusual telomeres of the viral genome, which are incompletely base-paired hairpins with an A+T content of >95 percent. Replication is robust and occurs from ~3 to 12 hours post-infection, leading to amounts of viral DNA that are estimated to approximate one-third the amount of the cellular DNA content.

Morphogenesis and Egress

As structural proteins and progeny become available, morphogenesis of nascent virions commences in the cytoplasm (Condit et al., 2006). Electron-dense areas of proteins destined for encapsidation are among the first hallmarks of morphogenesis. The appearance of membrane crescents (or cupules in three dimensions) at the periphery of these electron-dense areas is the first sign of membrane biogenesis. As these crescents enlarge and surround proteins destined to form the virion core, the curvature of a protein lattice that forms on their external face helps determine the size and shape of the immature virions (IV). Genome encapsidation is associated with the appearance of a nucleoid within these IV (forming IVN). Maturation of these IVN into infectious MV is accompanied by proteolytic processing of the major core proteins. This maturation is also accompanied by the transition of the oval IV to brick-shaped MV that have a characteristic dumbbell core.

The majority of the MV remain within the cell as long as the cell remains intact. A minority, however, are transported to the Golgi apparatus (or endosomal compartment), where they become wrapped with two additional lipid bilayers (forming wrapped virions, or WV). These outer envelopes contain a group of distinctive viral proteins that are not found in MV. WV traffic to the plasma membrane, where their outermost lipid bilayer fuses with the plasma membrane. This fusion leads to the exocytotic release of EV, which can disassociate from the plasma membrane and mediate distal spread. Alternatively, some EV remain bound to the plasma membrane at the site of egress. The viral proteins that were delivered to the subjacent plasma membrane during fusion activate the formation of actin tails, which then propel the attached EV toward neighboring cells, facilitating efficient proximal spread of the virus.

HOST SPECIFICITY AND RANGE

As noted earlier, humans are the sole host for variola and molluscum contagiosum viruses, and the success of the WHO-led smallpox eradication program was achievable because variola virus has no animal reservoir. In general, poxvirus infections in vertebrate hosts show species specificity; however, zoonotic infections do occasionally occur. The underlying mechanism for host tropism, which is determined largely by host–pathogen interaction at many levels, is not well understood (McFadden, 2005). The reservoir hosts for monkeypox are rodents and squirrels, but this virus can occasionally cross the species barrier to infect monkeys and humans (Di Giulio and Eckburg, 2004).

Poxviruses can bind to and enter a wide range of mammalian cells, but their success in replicating may vary. The ability of poxviruses to replicate and complete their viral life cycle in cells is dependent on many host-related factors, including cell type and species origin, cell cycle status, and intracellular signaling events leading to antiviral innate immunity and apoptosis. Identification of host range genes and elucidation of their interactions with host proteins and signaling pathways have shed light on poxvirus host tropism. For example, vaccinia E3L and K3L are host range genes targeting interferon (IFN)-inducible dsRNA-dependent protein kinase R (PKR) (Langland and Jacobs, 2002). E3L functions by sequestering dsRNA and preventing the activation of PKR, whereas K3L mimics eIF2α and acts as a pseudosubstrate for PKR, preventing phosphorylation of eIF2α and thus inhibition of protein synthesis. Suppression of PKR expression in a nonpermissive cell line for an E3L knockout virus (ΔE3L) restores viral protein synthesis and viral replication. Viral-induced apoptosis is blocked in PKR-deficient cells as well (Zhang et al., 2008). E3L also blocks the activation of another IFN-inducible protein, 2′–5′ oligoadenylate synthetase (2′–5′ OAS), and the subsequent activation of ribonuclease RNaseL (Rivas et al., 1998). The C-terminal dsRNA-binding domain of E3L is required for host range as well as pathogenesis. The N-terminal Z-DNA-binding domain of E3L is not required for host range, but it is required for pathogenesis in mice (Brandt and Jacobs, 2001).

Vaccinia K1L gene is an ankyrin-repeat containing host range protein that is involved in inhibiting IκBα degradation, which prevents activation of the host defense mechanisms of the NFkB pathway. The importance of the capacity to block the host cell response is demonstrated by the arrest of replication of vaccinia mutant strains lacking K1L at the stage of intermediate gene transcription in Chinese hamster ovary cells. The conservation of this function, and hence its likely importance to other orthopoxviruses, was shown by the rescue of this mutant by expression of another ankyrin-repeat containing host range protein from cowpox, CP77 (Ramsey-Ewing and Moss, 1996). Another ankyrin-repeat host range protein is M-T5 of myxoma virus, whose deletion leads to an inability to replicate in rabbit T lymphocytes and reduced virulence in European rabbits (Mossman et al., 1996). M-T5 is required for myxoma replication in certain human tumor cells (Sypula et al., 2004). Induction of IFN production has been shown to contribute to maintaining the species barrier for myxoma virus (Wang et al., 2004).

C7L is another host range gene, which may be functionally equivalent to K1L. Infection with vaccinia mutant lacking both K1L and C7L is nonpermissive in human and murine cells (Perkus et al., 1990; Oguiura et al., 1993). C7L homologues are present in the genomes of almost all mammallian poxviruses. A recent study by Meng and colleagues (2008) demonstrated that vaccinia C7L homologue, myxoma M62R, or yaba-like disease virus 67R, when reconstructed in a vaccinia mutant lacking K1L and C7L, restored the vaccinia mutant's ability to replicate in human and murine cells, possibly by suppressing PKR activation.

MVA and NYVAC are two candidates for third-generation vaccines against smallpox that have alterations in immunomodulatory and host range genes. MVA is derived from more than 500 serial passages of the vaccinia virus Ankara in chicken embryo fibroblasts (CEF); as a consequence, it lacks 31 kilobases of its genome (Antoine et al., 1998) (see Chapter 7). MVA does not replicate in most mammalian cells. The exceptions are BHK-21 cells and CEF. NYVAC is generated by deletion of 18 nonessential genes implicated in virulence or host range. It replicates in CEF, Vero cells, and BHK-21 cells (Tartaglia et al., 1992). MVA lacks K1L but retains C7L, whereas NYVAC lacks both K1L and C7L. In Hela cells, the MVA life cycle is blocked at a late stage of viral infection, probably at the assembly of immature virions (Sutter and Moss, 1992; Sancho et al., 2002). MVA with deletion of E3L still replicates in BHK-21 cells, but fails to replicate in CEF. Whereas MVA infection in Hela cells reveals a complete cascade of viral early, intermediate, and late gene transcription, MVA-ΔE3L infection produces only early and intermediate transcripts. It is related to the induction of 2′–5′ OAS/RNaseL and PKR (Ludwig et al., 2005, 2006). NYVAC induces apoptosis in Hela cells, which can be prevented through introduction of the C7L gene (Nájera et al., 2006).

HOST–PATHOGEN INTERACTIONS

Immune Modulation

Poxviruses, including variola, encode many genes that are known or predicted to modulate host antiviral responses (Seet et al., 2003). These include secreted viral proteins that bind cytokines, chemokines, and complement proteins, as well as intracellular antagonists that block key signaling pathways leading to establishment of an antiviral state, apoptosis, or proinflammatory responses. Over the last decade, significant advances have been made in the understanding of poxvirus–host interactions and viral immune modulatory genes.

Antiviral innate immunity is critical for the host to contain a viral infection initially and to activate the adaptive immune responses that result in viral clearance. Over the last several years, a number of viral-sensing pathways have been discovered in the host cell, including toll-like receptors (TLRs), retinoic acid-inducible gene I (RIG-I)-like receptors, nucleotide oligomerization domain (NOD)-like receptors (NLRs), and possibly cytosolic DNA sensors to detect viral nucleic acids and other components (Akira et al., 2006). The induction of type I IFN and proinflammatory cytokines and chemokines in various cell types in response to viral pathogens, including dendritic cells, macrophages, epithelial cells, and fibroblasts, leads to the further recruitment of other immune cells and the development of adaptive immunity. Both functional type I and type II IFN systems are required for protection against vaccinia infection. For example, mice that have targeted deletions in type I and type II IFN receptors show increased susceptibility to vaccinia infection (Müller et al., 1994). Although not studied for variola, IFNs can be predicted to be critical for the outcome of smallpox based on their fundamental contributions in other systemic viral infections.

Through interactions with their hosts, poxviruses have evolved to possess many mechanisms for antagonizing the production and actions of IFNs and proinflammatory cytokines and chemokines. Since the capacity of variola to block these responses is likely to be a key factor in its exceptional virulence for the human host, information about these mechanisms in other orthopoxviruses can be helpful in framing scientific questions about their contribution to the capacity of variola to overwhelm the human host. As an example of such mechanisms, many poxviruses encode IFNα/β-binding proteins and IFN-γ receptor homologs that dampen the effects of these molecules that are involved in countering viral replication and spread within the host (Upton et al., 1991; Symons et al., 1995). Specifically, the vaccinia virus (Western Reserve) B18R gene encodes for an IFNα/β-binding protein and is critical for virulence, as B18R knockout virus is attenuated in both murine intranasal and intracranial infection models (Symons et al., 1995). Similarly, ectromelia virus expresses functional homologs of IFN-γR and IFN-α/βR (Smith and Alcami, 2002). Recombinant ectromelia virus with deletion of the gene encoding type I IFN-binding protein was attenuated more than 107-fold compared with wild-type ectromelia virus (Xu et al., 2008). Variola virus also encodes homologs of IFN-γR and IFN-α/βR (Esposito et al., 2006; Li et al., 2007).

Tumor necrosis factor (TNF) is a pleiotropic cytokine that mediates inflammation and apoptosis within the host. Many poxviruses, including variola virus, encode TNF receptor homologs (TNFRs), which block the host TNF signaling pathway. Whether they do so by binding TNF or by preventing the oligomerization of TNFR through their preligand assembly domains (PLADs) remains controversial (Alejo et al., 2006; Sedger et al., 2006). Myxoma virus lacking the TNF receptor homolog M-T2 has decreased virulence compared with wild-type myxoma (Upton et al., 1991). Whereas cowpox encodes four viral TNFRs, including CrmB (cytokine response modifier B), CrmC, CrmD, and CrmE, variola virus encodes one CrmB-like protein. In addition to its ability to bind to TNF, the variola CrmB-like protein is capable of binding to everal chemokines (Alejo et al., 2006).

Poxviruses encode intracellular inhibitors that block caspase 1 activity, as well as IL-1β receptor homolog and IL-18-binding protein. IL-1β is another major cytokine mediating acute and chronic inflammation in response to infection and injury. It is produced through processing of its inactive precursor, proIL-1β, by caspase 1 to its active form, p17. Activation of caspase 1 is dependent on activation of a multimolecular complex termed an inflammasome, which can be triggered by various stimuli, including cytosolic bacterial infections, ATP, and alum. IL-18 is another potent inflammatory cytokine released as a result of the activation of caspase I. Vaccinia virus WR B15 encodes an IL-1β-binding protein. Infection of mice with a B15R deletion mutant induced fever, whereas wild-type vaccinia infection suppressed fever, a host response controlled by the presence of IL-1β (Alcami and Smith, 1996). Both variola virus and monkeypox encode CrmA that inhibits caspase 1, an IL-1β receptor, and an IL-18-binding protein (Seet et al., 2003). Myxoma virus attenuates inflammasome activation through an early viral gene, M13L, which interacts with a critical component of the inflammasome adapter protein, ASC (Johnston et al., 2005).

Poxviruses utilize multiple strategies to evade the chemokine system by encoding chemokine-binding proteins, chemokine receptor homologs, and chemokine receptor antagonists. Chemokines are a large family of small, secreted proteins that mediate the recruitment of immune cells to the sites of injury or infection. Chemokines can be divided into several categories depending on the number and position of the highly conserved cysteine residues near the N-terminus of the protein. The two major chemokine subfamilies are CC and CXC, and the two minor subfamilies are C and CX3C. Chemokines perform their functions by binding to their cognate seven transmembrane G-protein-coupled receptors on the surface of cells (Mantovani et al., 2006). Some chemokines have direct antiviral activity (Nakayama et al., 2006). Myxoma virus M-T7 encodes a low-affinity chemokine-binding protein that binds not only IFN-γ, but also a wide range of CXC, CC, and C chemokines. Myxoma M-T1 and vaccinia B29R encode high-affinity chemokine-binding proteins that bind to CC chemokine, but not to C, CXC, or CX3C chemokines (Graham et al., 1997; Alcami et al., 1998). Variola virus and monkeypox virus also have high-affinity chemokine-binding proteins (Alejo et al., 2006; Jones et al., 2008).

The complement system is required for successful host defense against poxviruses. Moulton and colleagues (2008) recently showed that mice deficient in complement are more susceptible to ectromelia (mousepox) infection. Variola, monkeypox, ectromelia, and vaccinia viruses encode complement-binding proteins that block both the classical and alternative pathways of complement activation (Kotwal and Moss, 1988; Rosengard et al., 2002; Liszewski et al., 2006; Parker et al., 2008).

Poxviruses also produce intracellular inhibitors to block the induction of type I IFN and proinflammatory cytokines and chemokines. Vaccinia virus A46R and A52R contain a Toll/IL-1 receptor (TIR) domain that can block the recruitment of adaptor molecules to IL-1R or TLR, preventing signaling following ligand–receptor interactions (Bowie et al., 2000). A52R has been found to block multiple TLRs through association with IRAK2 and TRAF6. Vaccinia mutant strain ΔA52R is attenuated in a murine intranasal model of infection (Harte et al., 2003). Recently, Vaccinia K7 was found to inhibit TBK1/IKKε-mediated IRF activation induced by TLR and non-TLR pathways (Schröder et al., 2008). Vaccinia E3L encodes a dsRNA-binding protein that can inhibit IRF3, IRF7, and NF-κB pathways, in addition to IFN-inducible genes (Smith et al., 2001; Xiang et al., 2002; Deng et al., 2006; Langland et al., 2006; Guerra et al., 2008) that would otherwise be activated following vaccinia infection. E3L also functions to subvert cytosolic RNA-sensing pathway mediated by MAVS and IRF3 in keratinocytes (DiPerna et al., 2004; Deng et al., 2008). Vaccinia N1L and K1L have been shown to block NF-κB signaling (Shisler and Jin, 2004). In addition, vaccinia VH1 gene encodes a phosphatase that blocks the activation of IFN-induced activation of STAT-1 (Najarro et al., 2001).

Apoptosis is an effective host mechanism for containing viral infections through programmed death of infected cells. Poxviruses have evolved to evade this defense mechanism through encoding of anti-apoptotic molecules. The IFN-inducible PKR promotes apoptosis through inhibition of protein synthesis, which can be counteracted by vaccinia E3L and K3L (Chang et al., 1992); variola virus encodes E3L and K3L homologs. Myxoma M11L, a virulence gene, encodes a mitochondria-targeted molecule that prevents apoptosis (Everett et al., 2000), while molluscum contagiosum virus encodes two genes, MC159 and MC160, to block the activation of initiator caspase, caspase 8 (Shisler and Moss, 2001). CrmA, a member of the serine protease inhibitor family, first identified as an inhibitor for IL-1β-converting enzyme (caspase 1), also inhibits caspase 8 and blocks apoptosis induced by various factors (Tewari and Dixit, 1995).

Adaptive Immune Response

Smallpox was eradicated prior to the development of modern quantitative cellular assays that measure virus-specific T cell numbers and function. Likewise, because of the conditions associated with smallpox outbreaks, monitoring humoral immune responses was usually difficult. Although neutralizing antibodies were measured, other more rapid and quantitative approaches, such as enzyme-linked immunosorbent assays (ELISAs), were not well established. As a result, knowledge about smallpox immunity is limited, and concepts of smallpox immunity are based on indirect information that is available about the kinetics, magnitude, and duration of orthopoxvirus-specific immunity derived from analysis of vaccinia-specific T cell and antibody responses elicited by immunization with vaccinia virus. Since vaccinia replication is well controlled while variola is often life-threatening, how well the host response to vaccinia mimics that induced by variola is not clear (see also Chapter 7). Nevertheless, these vaccinia responses define a protective cross-reactive response that is relevant for vaccine design and may suggest the characteristics of those responses that protected smallpox survivors from reinfection. Following vaccinia inoculation, antiviral T cell responses are difficult to detect at 1 week, but then rapidly expand and peak within approximately 2 weeks of infection (Miller et al., 2008). Antiviral antibody responses are slightly delayed in comparison with the T cell responses and generally peak within 2–3 weeks of inoculation. After the vaccinia lesion at the site of inoculation has resolved, antiviral T cell and antibody responses decline rapidly before reaching a more long-lived plateau phase in which immunological memory is maintained for decades. During this memory phase, vaccinia-specific $CD4^+$ and $CD8^+$ T cell responses decline slowly, with an estimated half-life of 8–15 years (Crotty et al., 2003; Hammarlund et al., 2003), whereas antiviral antibody responses are more stable, with an estimated half-life of 92 years (Amanna et al., 2007).

No specific immunological correlate predicts protection against smallpox or any of the other orthopoxviruses. Animal model experiments indicate that the induction of immunity to proteins present in the EV form is essential for complete protection against challenge (Kaufman et al., 2008). Results of studies in nonhuman primates indicate that vaccine-mediated immunity against lethal monkeypox challenge is due to the presence of neutralizing antibodies (Edghill-Smith et al., 2005). Earlier studies in humans also demonstrated that smallpox patients who developed higher antibody responses during acute smallpox infection had a lower mortality rate than those who mounted weaker antibody responses (Slifka, 2004). However, these individuals could also have had a poor virus-specific T cell response. Administration of high-dose convalescent serum to smallpox patients appeared to protect against lethal smallpox infection in uncontrolled clinical studies (Slifka, 2004). It is likely that preexisting antiviral antibodies provide a first line of protection against infection with orthopoxviruses, whereas antiviral T cells, along with inhibitory antibodies, are needed if the virus overcomes this barrier and gains entry into the host. During primary infection, effective induction of both T cell and B cell responses is probably necessary to prevent the virus from causing fatal complications and to clear infectious virus from the host (Slifka, 2004).

REFERENCES

Akira, S., S. Uematsu, and O. Takeuchi. 2006. Pathogen recognition and innate immunity. *Cell* 124:783–801.

Alcami, A., and G. L. Smith. 1996. A mechanism for the inhibition of fever by a virus. *Proceedings of the National Academy of Sciences of the United States of America* 93:11029–11034.

Alcami, A., J. A. Symons, P. D. Collins, T. J. Williams, and G. L. Smith. 1998. Blockade of chemokine activity by a soluble chemokine binding protein from vaccinia virus. *Journal of Immunology* 160:624–633.

Alejo, A., M. B. Ruiz-Argüello, Y. Ho, V. P. Smith, M. Saraiva, and A. Alcami. 2006. A chemokine-binding domain in the tumor necrosis factor receptor from variola (smallpox) virus. *Proceedings of the National Academy of Sciences of the United States of America* 103:5995–6000.

Amanna, I. J., N. E. Carlson, and M. K. Slifka. 2007. Duration of humoral immunity to common viral and vaccine antigens. *New England Journal of Medicine* 357:1903–1915.

Antoine, G., F. Scheiflinger, F. Dorner, and F. G. Falkner. 1998. The complete genomic sequence of the modified vaccinia Ankara strain: Comparison with other orthopoxviruses. *Virology* 244(2):365–396.

Biel, S. S., and H. R. Gelderblom. 1999. Diagnostic electron microscopy is still a timely and rewarding method. *Journal of Clinical Virology* 13:105–119.

Bowie, A., E. Kiss-Toth, J. A. Symons, G. L. Smith, S. K. Dower, and L. A. J. O'Neill. 2000. A46R and A52R from vaccinia virus are antagonists of host IL-1 and toll-like receptor signaling. *Proceedings of the National Academy of Sciences of the United States of America* 97:10162–10167.

Brandt, T. A., and B. L. Jacobs. 2001. Both carboxyl- and amino-terminal domains of the vaccinia virus interferon resistance gene, E3L, are required for pathogenesis in a mouse model. *Journal of Virology* 75:850–856.

Broyles, S. S. 2003. Vaccinia virus transcription. *Journal of General Virology* 84:2293–2303.

Carter, G. C., M. Law, M. Hollinshead, and G. L. Smith. 2005. Entry of the vaccinia virus intracellular mature virion and its interactions with glycosaminoglycans. *Journal of General Virology* 86:1279–1290.

Chang, H. W., J. C. Watson, and B. L. Jacobs. 1992. The E3L gene of vaccinia virus encodes an inhibitor of the interferon-induced, double-stranded RNA-dependent protein kinase. *Proceedings of the National Academy of Sciences of the United States of America* 89(11):4825–4829.

Chen, N., G. Li, M. K. Liszewski, J. P. Atkinson, P. B. Jahrling, Z. Feng, J. Schriewer, C. Buck, C. Wang, E. J. Lefkowitz, J. J. Esposito, T. Harms, I. K. Damon, R. L. Roper, C. Upton, and R. M. Buller. 2005. Virulence differences between monkeypox virus isolates from West Africa and the Congo basin. *Virology* 340:46–63.

Chiu, W. L., C. L. Lin, M. H. Yang, D. L. Tzou, and W. Chang. 2007. Vaccinia virus 4c (A26L) protein on intracellular mature virus binds to the extracellular cellular matrix laminin. *Journal of Virology* 81:2149–2157.

Chung, C. S., J. C. Hsiao, Y. S. Chang, and W. Chang. 1998. A27L protein mediates vaccinia virus interaction with cell surface heparan sulfate. *Journal of Virology* 72:1577–1585.

Condit, R. C. 2007. Vaccinia, Inc.—Probing the functional substructure of poxviral replication factories. *Cell Host and Microbe* 2:205–207.

Condit, R. C., N. Moussatche, and P. Traktman. 2006. In a nutshell: Structure and assembly of the vaccinia virion. *Advances in Virus Research* 66:31–124.

Crotty, S., P. Felgner, H. Davies, J. Glidewell, L. Villarreal, and R. Ahmed. 2003. Cutting edge: Long-term B cell memory in humans after smallpox vaccination. *Journal of Immunology* 171:4969–4973.

Deng, L., P. Dai, W. Ding, R. D. Granstein, and S. Shuman. 2006. Vaccinia virus infection attenuates innate immune responses and antigen presentation by epidermal dendritic cells. *Journal of Virology* 80:9977–9987.

Deng, L., P. Dai, T. Parikh, H. Cao, V. Bhjoy, Q. Sun, Z. J. Chen, T. Merghoub, A. Houghton, and S. Shuman. 2008. Vaccinia virus subverts a mitochondrial antiviral signaling protein-dependent innate immune response in keratinocytes through its dsRNA binding protein E3. *Journal of Virology* 82:10735–10746.

Di Giulio, D. B., and P. B. Eckburg. 2004. Human Monkeypox: An emerging zoonosis. *The Lancet Infectious Diseases* 4(1):15–25.

DiPerna, G., J. Stack, A. G. Bowie, A. Boyd, G. Kotwal, A. Zhang, S. Arvikar, E. Latz, K. A. Fitzgerald, and W. L. Marshall. 2004. Poxvirus protein N1L targets the I-κB kinase complex, inhibits signaling to NF-κB by the tumor necrosis factor superfamily of receptors, and inhibits NF-κB and IRF3 signaling by toll-like receptors. *The Journal of Biological Chemistry* 279:36570–36578.

Edghill-Smith, Y., H. Golding, J. Manischewitz, L. R. King, D. Scott, M. Bray, A. Nalca, J. W. Hooper, C. A. Whitehouse, J. E. Schmitz, K. A. Reimann, and G. Franchini. 2005. Smallpox vaccine-induced antibodies are necessary and sufficient for protection against monkeypox virus. *Nature Medicine* 11:740–747.

Esposito, J. J., S. A. Sammons, A. M. Frace, J. D. Osborne, M. Olsen-Rasmussen, M. Zhang, D. Govil, I. K. Damon, R. Kline, M. Laker, Y. Li, G. L. Smith, H. Meyer, J. W. LeDuc, and R. M. Wohlhueter. 2006. Genome sequence diversity and clues to the evolution of variola (smallpox) virus. *Science* 313(5788)807–812.

Everett, H., M. Barry, S. F. Lee, X. Sun, K. Granham, J. Stone, R. C. Bleakley, and G. McFadden. 2000. M11: A novel mitochondria-localized protein of myxoma virus that blocks apoptosis of infected leukocytes. *Journal of Experimental Medicine* 191(9): 1487-1498.

Garon, C.F., Barbosa, E., and Moss B. 1978. Visualization of an inverted terminal repetition in vaccinia virus DNA. *Proceedings of the National Academy of Sciences of the United States of America* 75(10):4863-4867.

Graham, K. A., A. S. Lalani, J. L. Macen, T. L. Ness, M. Barry, L. Y. Liu, A. Lucas, I. Clark-Lewis, R. W. Moyer, and G. McFadden. 1997. The T1/35kDa family of poxvirus-secreted proteins bind chemokines and modulate leukocyte influx into virus–infected tissues. *Virology* 229:12–24.

Guerra, S., A. Cáceres, K.-P. Knobeloch, I. Horak, and M. Esteban. 2008. Vaccinia virus E3 protein prevents the antiviral action of ISG15. *PLos Pathogen* 4:e1000096.

Hammarlund, E., M. W. Lewis, S. G. Hansen, L. I. Strelow, J. A. Nelson, G. J. Sexton, J. M. Hanifin, and M. K. Slifka. 2003. Duration of antiviral immunity after smallpox vaccination. *Nature Medicine* 9:1131–1137.

Harte, M. T., I. R. Haga, G. Maloney, P. Gray, P. C. Reading, N. W. Bartlett, G. L. Smith, A. Bowie, and L. A. J. O'Neill. 2003. The poxvirus protein A52R targets toll-like receptor signaling to suppress host defense. *Journal of Experimental Medicine* 197:343–351.

Johnston, J. B., J. W. Barrett, SH Nazarian, M. Goodwin, D. Ricuttio, G. Wang, and G. McFadden. 2005. A poxvirus-encoded pyrin domain protein interacts with ASC-1 to inhibit host inflammatory and apoptotic responses to infection. *Immunity* 23:587–598.

Jones, J. M., I. Messauodi, R. D. Estep, B. Orzechowska, and S.W. Wong. 2008. Monkeypox virus viral chemokine inhibitor (MPV vCCI), a potent inhibitor of rhesus macrophage inflammatory protein-1. *Cytokine* 43:220-228.

Katsafanas, G. C., and B. Moss. 2007. Colocalization of transcription and translation within cytoplasmic poxvirus factories coordinates viral expression and subjugates host functions. *Cell Host and Microbe* 2:221–228.

Kaufman, D. R., J. Goudsmit, L. Holterman, B. A. Ewald, M. Denholtz, C. Devoy, A. Giri, L. E. Grandpre, J. M. Heraud, G. Franchini, M. S. Seaman, M. J. Havenga, and D. H. Barouch. 2008. Differential antigen requirements for protection against systemic and intranasal vaccinia virus challenges in mice. *Journal of Virology* 82(14):6829–6837.

Kotwal, G. J., and B. Moss. 1988. Vaccinia virus encodes a secretory polypeptide structurally related to complement control proteins. *Nature* 335:176-178.

Langland, J. O., and B. L. Jacobs. 2002. The role of the PKR-inhibitory genes, E3L and K3L, in determining vaccinia virus host range. *Virology* 299:133–141.

Langland, J. O., J. C. Kash, V. Carter, M. J. Thomas, M. G. Katze, and B. L. Jacobs. 2006. Suppression of proinflammatory signal transduction and gene expression by the dual nucleic acid binding domains of the vaccinia virus E3L proteins. *Journal of Virology* 80:10083–10095.

Lefkowitz, E. J., C. Wang, and C. Upton. 2006. Poxviruses: Past, present and future. *Virus Research* 117:105–118.

Li, Y., D. S. Carroll, S. N. Gardner, M. C. Walsh, E. A. Vitalis, and I. K. Damon. 2007. On the origin of smallpox: Correlating variola phylogenics with historical smallpox records. *Proceedings of the National Academy of Sciences of the United States of America* 104(40):15787–15792.

Liszewski, M. K., Leung, M. K., Hauhart, R., Buller, R. M., Bertram, P., Wang, X., Rosengard, A. M., Kotwal, G. J., and J.P. Atkinson. 2006. Structure and regulatory profile of the monkeypox inhibitor of complement: comparison to homologs in vaccinia and variola and evidence for dimer formation. *Journal of Immunology* 176: 3725–3734.

Ludwig, H., J. Mages, C. Staib, M. H. Lehmann, R. Lang, and G. Sutter. 2005. Role of viral factor E3L in modified vaccinia virus ankara infection of human HeLa cells: Regulation of the virus life cycle and identification of differentially expressed host genes. *Journal of Virology* 79(4):2584–2596.

Ludwig, H., Y. Suezer, Z. Waibler, U. Kalinke, B. S. Schnierle, and G. Sutter. 2006. Double-stranded RNA-binding protein E3 controls translation of viral intermediate RNA, marking an essential step in the life cycle of modified vaccinia virus Ankara. *The Journal of General Virology* 87(Pt. 5):1145–1155.

Mantovani, A., R. Bonecchi, and M. Locati. 2006. Tuning inflammation and immunity by chemokine sequestration: Decoys and more. *Nature Reviews Immunology* 6:907–918.

McFadden, G. 2005. Poxvirus tropism. *Nature Reviews. Microbiology* 3(3):201–213.

Mercer, J., and A. Helenius. 2008. Vaccinia virus uses macropinocytosis and apoptotic mimicry to enter host cells. *Science* 320:531–535.

Meng, X., J. Chao, and Y. Xiang. 2008. Identification from diverse mammalian poxviruses of host-range regulatory genes functioning equivalently to vaccinia virus C7L. *Virology* 372(2):372–383.

Miller, J. D., R. G. van der Most, R. S. Akondy, J. T. Glidewell, S. Albott, D. Masopust, K. Murali-Krishna, P. L. Mahar, S. Edupuganti, S. Lalor, S. Germon, C. Del Rio, M. J. Mulligan, S. I. Staprans, J. D. Altman, M. B. Feinberg, and R. Ahmed. 2008. Human effector and memory CD8+ T cell responses to smallpox and yellow fever vaccines. *Immunity* 28:710–722.

Moss, B. 2006. Poxvirus entry and membrane fusion. *Virology* 344:48–54.

Moss, B. M. 2007. Poxviridae: The viruses and their replication. In *Fields' virology,* 5th edition, edited by B. N. Fields, D. M. Knipe, P. M. Howley, and D. E. Griffin. Philadelphia, PA: Lippincott Williams & Wilkins. Pp. 2906–2946.

Moss, B., and F. S. De Silva. 2006. Poxvirus. In *DNA replication and human disease*, edited by M. DePamphilis. Cold Spring Harbor, NY: Cold Spring Harbor Laboratory Press. Pp. 707–727.

Mossman, K., S. F. Lee, M. Barry, L. Boshkov, and G. McFadden. 1996. Disruption of M-T5, a novel myxoma virus gene member of the poxvirus host range superfamily, results in dramatic attenuation of myxomatosis in infected European rabbits. *Journal of Virology* 70:4394–4410.

Moulton, E. A., J. P. Atkinson, and R. M. Buller. 2008. Surviving mousepox infection requires the complement system. *PLos Pathogen* 4(12):e1000249.

Müller, U., U. Steinhoff, L. F. Reis, S. Hemmi, J. Pavlovic, R. M. Zinkernagel, and M. Aguet. 1994. Functional role of type I and type II interferons in antiviral defense. *Science* 264:1918–1921.

Najarro, P., P. Traktman, and J. A. Lewis. 2001. Vaccinia virus blocks gamma interferon signal transduction: Viral VH1 phosphatase reverses Stat1 activation. *Journal of Virology* 75:3185–3196.

Nájera, J. L., C. E. Gómez, E. Domingo-Gil, M. M. Gherardi, and M. Esteban. 2006. Cellular and biochemical differences between two attenuated poxvirus vaccine candidates (MVA and NYVAC) and role of the C7L gene. *Journal of Virology* 80(12):6033–6047.

Nakayama, T., J. Shirane, K. Hieshima, M. Shibano, M. Watanabe, Z. Jin, D. Nagakubo, T. Saito, Y. Shimomura, and O. Yoshie. 2006. Novel antiviral activity of chemokines. *Virology* 350:484–492.

Oguiura, N., D. Spehner, and R. Drillien. 1993. Detection of a protein encoded by the vaccinia virus C7L open reading frame and study of its effect on virus multiplication in different cell lines. *The Journal of General Virology* 74(Pt. 7):1409–1413.

Parker, A. K., W. M. Yokoyama, J. A. Corbett, N. Chen, and R. M. Buller. 2008. Primary naive and interleukin-2-activated natural killer cells do not support efficient ectromelia virus replication. *Journal of General Virology* 89:751-759.

Parrish, S., and B. Moss. 2007. Characterization of a second vaccinia virus mRNA-decapping enzyme conserved in poxviruses. *Journal of Virology* 81:12973–12978.

Parrish, S., W. Resch, and B. Moss. 2007. Vaccinia virus D10 protein has mRNA decapping activity, providing a mechanism for control of host and viral gene expression. *Proceedings of the National Academy of Sciences of the United States of America* 104:2139–2144.

Perkus, M. E., S. J. Goebel, S. W. Davis, G. P. Johnson, K. Limbach, E. K. Norton, and E. Paoletti. 1990. Vaccinia virus host range genes. *Virology* 179(1):276–286.

Ramsey-Ewing, A. L., and B. Moss. 1996. Complementation of a vaccinia virus host-range K1L gene deletion by the non-homologous CP77 gene. *Virology* 222:75–86.

Reed, K. D., J. W. Melski, M. B. Graham, R. L. Regnery, M. J. Sotir, M. V. Wegner, J. J. Kazmierczak, E. J. Stratman, Y. Li, J. A. Fairley, G. R. Swain, V. A. Olson, E. K. Sargent, S. C. Kehl, M. A. Frace, R. Kline, S. L. Foldy, J. P. Davis, and I. K. Damon. 2004. The detection of monkeypox in humans in the Western Hemisphere. *New England Journal of Medicine* 350:342–350.

Rivas, C., J. Gil, Z. Mělková, M. Esteban, and M. Díaz-Guerra. 1998. Vaccinia virus E3L protein is an inhibitor of the interferon(i.f.n.)-induced 2-5A synthetase enzyme. *Virology* 1243:406–414.

Rosengard, A.M., Liu, Y., Nie, Z., and R. Jimenez. 2002. Variola virus immune evasion design: expression of a highly efficient inhibitor of human complement. *Proceedings of the National Academy of Sciences of the United States of America* 99:8808–8813.

Sancho, M. C., S. Schleich, G. Griffiths, and J. Krijnse-Locker. 2002. The block in assembly of modified vaccinia virus Ankara in HeLa cells reveals new insights into vaccinia virus morphogenesis. *Journal of Virology* 76(16):8318–8334.

Schröder, M., M. Barab, and A. G. Bowie. 2008. Viral targeting of DEAD box protein 3 reveals its role in TBK1/IKKε-mediated IRF activation. *EMBO* 27:2147–2157.

Sedger, L. M., S. R. Osvath, X. M. Xu, G. Li, F. K. Chan, J. W. Barrett, and G. McFadden. 2006. Poxvirus tumor necrosis factor receptor (TNFR)-like T2 proteins contain a conserved preligand assembly domain that inhibits cellular TNFR1-induced cell death. *Journal of Virology* 80(18):9300–9309.

Seet, B. T., J. B. Johnston, C. R. Brunetti, J. W. Barrett, H. Everett, C. Cameron, J. Sypula, S. H. Nazarian, A. Lucas, and G. McFadden. 2003. Poxviruses and immune evasion. *Annual Review of Immunology* 21:377–423.

Shisler, J. L., and X. L. Jin. 2004. The vaccinia virus K1L gene product inhibits host NF-κB activation by preventing IκBα degradation. *Journal of Virology* 78:3553–3560.

Shisler, J.L., and B. Moss. 2001. Molluscum contagiosum virus inhibitors of apoptosis: The MC159 v-FLIP protein blocks Fas-induced activation of procaspases and degradation of the related MC160 protein. *Virology* 282:14–25.

Slifka, M. K. 2004. Immunological memory to viral infection. *Current Opinion in Immunology* 16:443–450.

Smith, E. J., I. Marié, A. Prakash, A. García-Sastre, and D. E. Levy. 2001. IRF3 and IRF7 phosphorylation in virus-infected cells does not require double-stranded RNA-dependent protein kinase R or Ikappa B kinase but is blocked by Vaccinia virus E3L protein. *Journal of Biological Chemistry* 276:8951–8957.

Smith, G. L. and Law, M. 2004. The Exit of Vaccinia Virus From Infected Cells. *Virus Research* 106(2): 189–197.

Smith, V. P., and A. Alcami. 2002. Inhibition of interferons by ectromelia virus. *Journal of Virology* 76:1124–1134.

Sutter, G., and B. Moss. 1992. Nonreplicating vaccinia vector efficiently expresses recombinant genes. *Proceedings of the National Academy of Sciences of the United States of America* 89(22):10847–10851.

Symons, J. A., A. Alcami, and G. L. Smith. 1995. Vaccinai virus encodes a soluble type I interferon receptor of novel structure and broad species specificity. *Cell* 81:551–560.

Sypula, J., F. Wang, Y. Ma, J. Bell, and G. McFadden. 2004. Myxoma virus tropism in human tumor cells. *Gene Therapy & Molecular Biology* 8:103–114.

Tartaglia, J., M. E. Perkus, J. Taylor, E. K. Norton, J. C. Audonnet, W. I. Cox, S. W. Davis, J. van der Hoeven, B. Meignier, and M. Riviere. 1992. NYVAC: A highly attenuated strain of vaccinia virus. *Virology* 188(1):217–232.

Tewari, M., and V. M. Dixit. 1995. Fas- and tumor necrosis factor-induced apoptosis is inhibited by the poxvirus crmA gene product. *Journal of Biological Chemistry* 270:3255–3260.

Townsley, A. C., A. S. Weisberg, T. R. Wagenaar, and B. Moss. 2006. Vaccinia virus entry into cells via a low-pH-dependent endosomal pathway. *Journal of Virology* 80:8899–8908.

Upton, C., J. L. Macen, M. Schreiber, and G. McFadden. 1991. Myxoma virus expresses a secreted protein with homology to the tumor necrosis factor receptor gene family that contributes a viral virulence. *Virology* 184:370–382.

Upton, C., S. Slack, A. L. Hunter, A. Ehlers, and R. L. Roper. 2003. Poxvirus orthologous clusters: Toward defining the minimum essential poxvirus genome. *Journal of Virology* 77:7590–7600.

Wang, F., Y. Ma, J. W. Barrett, X. Gao, J. Loh, E. Barton, H. W. Virgin, and G. McFadden. 2004. Disruption of ERK-dependent type I interferon induction breaks the myxoma virus species barrier. *Nature Immunology* 5:1266–1274.

Xiang, Y., R. C. Condit, S. Vijaysri, B. L. Jacobs, B. R. Williams, and R. H. Silverman. 2002. Blockade of interferon induction and action by the E3L double-stranded RNA binding proteins of vaccinia virus. *Journal of Virology* 76:5251–5259.

Xu, R.-H., M. Cohen, Y. Tang, E. Lazear, J. C. Whitbeck, R J. Eisenberg, G. H. Cohen, and L. J. Sigal. 2008. The orthopoxvirus type I IFN binding protein is essential for virulence and an effective target for vaccination. *Journal of Experimental Medicine* 205:981–992.

Zhang, P., B. L. Jacobs, and C. E. Samuel. 2008. Loss of protein kinase PKR expression in human Hela cells complements the vaccinia virus E3L deletion mutant phenotype by restoration of viral protein synthesis. *Journal of Virology* 82:840–848.

4

Animal Models Using Variola and Other Orthopoxviruses

In 1999, there were no suitable animal models for variola. This led the IOM committee at that time to draw the following conclusion:

> **The existence of animal models would greatly assist the development and testing of antiviral agents and vaccines, as well as studies of variola pathogenesis. Such a program could be carried out only with live variola virus.**

Since 1999, some progress has been made in developing animal models. However, it should be emphasized that there is still no animal model that satisfactorily recapitulates all relevant aspects of human smallpox. Although the nonhuman primate model described by Jahrling and colleagues (2004) offers some features that are suggestive of later-stage, fulminant human smallpox, the ability of this model system to mimic the wide spectrum of human disease manifestations and pathophysiology remains uncertain. This chapter describes efforts to use animal models to study variola infection and disease in humans. Although such efforts have included use of the variola virus, the inability to infect most animal species with variola has resulted in attempts to use vaccinia, cowpox, and mousepox; monkeypox; and myxoma virus.

VARIOLA

As discussed in Chapter 3, there are no animal reservoirs for variola virus in nature, and most animal species cannot be infected in the laboratory. Attempts to develop nonhuman primate models for variola infection and disease in the 1960s met with only limited success (Hahon and Wilson, 1960; Hahon, 1961; Lancaster et al., 1966; Westwood et al., 1966). Although these attempts reportedly produced rash and systemic illness in various nonhuman primate species, disease was inconsistent and variable in severity with each of the routes of inoculation and host species examined. Of note, the reports described primary and secondary viremias and a mild, brief illness with fever and rash in *Macaca irus* and *Macaca mulatta* exposed to variola virus via the aerosol route (Hahon and Wilson, 1960; Lancaster et al., 1966; Westwood et al., 1966).

Motivated by the potential utility of a realistic and consistent model with which to test drugs and vaccines intended for possible use against smallpox in humans, CDC revisited the possibility of developing a model of human smallpox in nonhuman primates at the beginning of this decade (Jahrling et al., 2004). After limited unsuccessful efforts to produce consistent, severe disease with an aerosol device, the investigators turned to an intravenous route of infection in cynomolgus macaques. Lethal disease with systemic features reminiscent of late-stage severe human smallpox, including skin vesicles and pustules, was achieved with doses of 10^9 plaque-forming units. Lesser doses, as low as 10^6 plaque-forming units, produced less severe disease and fewer skin lesions in a dose-dependent but less consistent manner. As described, this macaque model appears to truncate the natural course of variola infection, bypassing the early respiratory tract replication of the virus, primary viremia, and early clinical phases of human smallpox. On the other hand, aspects of the systemic pathology resembled some of the reported features of lethal, hemorrhagic smallpox in humans as described in the historical record. But because the route of infection and the dose differed substantially from those of the natural setting, there is reason to believe that the mechanisms of pathogenesis in this model may vary from those that took place during natural disease. For example, the initial instantaneous viremia preempts a prodromal period and circumvents early local replication of virus in the respiratory tract. An adequate assessment of these issues has not been possible, in part because human smallpox was eradicated before modern investigatory tools became available and also because relatively few studies of the current nonhuman primate model have been published.

Nonetheless, this nonhuman primate model was subsequently used to test therapeutics and vaccines (see Chapters 6 and 7, respectively), with the understanding that it sets a stringent, perhaps overly demanding standard for efficacy. Proposed improvements of this macaque model include using an intratracheal route of inoculation to achieve a consistent and more realistic course of disease.

Elucidation of features of variola pathogenesis, a secondary goal in the development of this macaque model, has been achieved using traditional and more modern tools (Rubins et al., 2004). In addition to a variety of histology-, hematology-, chemistry-, and immunology-based measurements, genome-wide features of nonhuman primate gene expression were measured in serial peripheral blood specimens using DNA microarrays. The latter revealed gene transcript abundance patterns indicative of prominent interferon and cell proliferation responses, and notable for the absence of responses associated with tumor necrosis factor alpha and transcription factor NF-kappaB, which would otherwise be typical of many acute overwhelming infections. These and other findings enhance understanding of the mechanisms responsible for variola-associated morbidity and mortality, as well as possible new targets for therapeutic intervention.

VACCINIA, COWPOX, AND MOUSEPOX

Vaccinia virus is perhaps the most widely used poxvirus in animal models for studying variola virus infection because of its ready availability and extensive knowledge base, and the susceptibility of laboratory rabbits and mice to vaccinia infection. In rabbits in particular, infection with a rabbit-adapted strain of vaccinia virus—rabbitpox virus—generates a disease that recapitulates some of the important features of smallpox, including transmission between hosts by the aerosol route and a generalized rash (Adams et al., 2007).

Laboratory mice can be lethally infected with several strains of vaccinia virus introduced intranasally or by aerosol. This mouse model yields dose-dependent lethality, with up to 100 percent of animals dying, and can be refined to result in a sublethal disease course in which disease severity is quantified by weight loss followed by recovery over a period of 2–3 weeks. Although this model is characterized by a more rapid disease onset—around 3 days—than is seen with smallpox and is not characterized by a rash, it has been used to compare the efficacy of novel and traditional vaccines and to conduct research on antiviral therapies (Bray et al., 2000; Smee et al., 2001; Belyakov et al., 2003; Hooper et al., 2003; McCurdy et al., 2004; Wyatt et al., 2004; Law et al., 2005; Phelps et al., 2005, 2007; Abdalrhman et al., 2006; Ferrier-Rembert et al., 2007).

Inbred mice are also useful in models for cowpox (Bray et al., 2000; Ferrier-Rembert et al., 2007) and mousepox (Fenner, 1949) infection, and in both cases, a lethal challenge is obtainable. The mouse/cowpox model is broadly similar to that of mouse/vaccinia, and has value in extending the range of orthopoxviruses that can be used in a single host for the evaluation of measures that may control orthopoxvirus infection. The mouse/mousepox model is somewhat different in that it is lethal at very low doses and is restricted to a single host, and its severity can be viewed as generating a model that is more relevant to variola infection in humans. However, there are significant differences between the disease course of mousepox in mice and most other acute orthopoxvirus infections. Mousepox is characterized by large ulcerating lesions rather than the discrete maculo-papular rash characteristic of smallpox, and extensive liver damage is seen as well (Jones et al., 1997). Although the disease severity and host restriction make the mousepox model attractive for testing control measures, this model must be treated with caution given this differential pathology. However, the ability to perform challenge experiments with vaccinia, cowpox, and mousepox viruses in a single species when protection against all three viruses can be achieved with traditional smallpox vaccine adds considerably to the confidence with which extrapolations from these models to human smallpox can be made.

MONKEYPOX

Monkeypox is perhaps the most relevant orthopoxvirus with regard to nonvariola animal models for smallpox. Monkeypox virus causes a sometimes fatal disease in humans whose clinical features and course are similar to those of smallpox. The use of this virus in the laboratory requires biosafety level (BSL)-3 conditions in the United States, although the disease is prevented in humans and animals with smallpox vaccine. The classification of monkeypox virus as a select agent in the United States further complicates and hinders work on this virus.

As its name implies, monkeypox virus causes disease in nonhuman primates, and it has been used experimentally to cause disease in macaques that is similar to smallpox and monkeypox in humans, and to evaluate possible countermeasures against smallpox (Earl et al., 2004; Stittelaar et al., 2005). Monkeypox does not cause significant disease in laboratory mice; however, the recent discovery that it causes disease in North American prairie dogs has led to the examination of other, related ground-dwelling squirrels, and there are now several rodent-based models for orthopoxvirus disease using this virus (Tesh et al., 2004; Hutson et al., 2007).

The most important feature of monkeypox is not its similarity in humans to smallpox. Rather, monkeypox is a public health problem in its own right. At least 88 cases with 3 fatalities occur annually in endemic regions of central Africa (Hutin et al., 2001, Levine et al., 2007; Parker et al., 2007). Public health issues pertaining to monkeypox are beyond the scope of this study. Nevertheless, the committee notes that human monkeypox deserves attention because of its toll in endemic areas, and the licensure of therapies for the disease would provide a tangible benefit for a large at-risk population. Moreover, lessons learned from the development of licensed medical countermeasures for human monkeypox might address many of the uncertainties associated with extrapolation among different orthopoxviruses in animal models.

MYXOMA VIRUS

Animal models with relevance to smallpox are generally restricted to orthopoxviruses—members of the same genus as variola itself. One nonorthopoxvirus is worthy of consideration, however, because of its pathogenicity and the body of research based on its use. Myxoma virus, a poxvirus of the leporipoxvirus genus, is a virus of New World rabbits of the genus *Sylvilagus*, in which it causes an infection that is almost asymptomatic and is nonlethal. When introduced to European *Oryctolagus* rabbits, myxoma virus causes a fulminant ulcerating infection known as myxomatosis, with a very high mortality rate (Stanford et al., 2007). The severity of disease in myxoma virus-infected European rabbits invites parallels with smallpox, and although the two viruses differ significantly in pathology, so, too, do smallpox and models using vaccinia, cowpox, and mousepox.

Like orthopoxviruses, myxoma virus produces a number of proteins that interact with elements of the immune system. Many of the lessons learned from studies with myxoma virus directly inform and influence understanding of orthopoxviruses and vice versa. However, myxoma virus is sufficiently different from orthopoxviruses that smallpox vaccine does not protect European rabbits from myxomatosis, and ST-246, a promising candidate antiviral drug for treatment of orthopoxvirus diseases, including smallpox, has no activity against myxoma virus, which lacks the specific target of the drug in orthopoxviruses. Consequently, data from animal models using myxoma virus cannot be extrapolated to smallpox in humans.

CHIMERIC VIRUSES

While the extreme host restriction of variola virus greatly facilitated the smallpox eradication campaign, it also hampered research because no animal model using variola was available. In the 1960s, attempts were made to address this gap by constructing chimeric viruses from variola and either cowpox or rabbitpox (a rabbit-adapted strain of vaccinia virus) viruses (Bedson and Dumbell, 1964a,b). These chimeric viruses were constructed by coinfection of cell lines with the two viruses and plaque purification of random recombinants between the two. At the time, the random nature of the resulting recombinant viruses and the inability to fully sequence these recombinants restricted their utility for research into the pathogenesis of variola infection. Moreover, the advisability of adapting a human-only virus to growth in animals that could thereby become potential reservoir hosts was questioned. These chimeric viruses, which are stored under BSL-4 containment at CDC, were generated with methods that yield random

TABLE 4-1 Usefulness of Animal Models and Human Infections for Understanding the Pathogenesis of Variola and Other Orthopoxviruses and for Developing Therapeutics and Vaccines

	Small Animal	Nonhuman Primate	Human
Orthopoxviruses other than monkeypox and variola	Contribution to overall understanding of pathogenesis of infections caused by poxviruses; provides some evidence of antiviral activity of drugs against poxviruses related to variola, but not variola.	Limited usefulness: some potential to help in identifying useful interventions against smallpox in humans,[a] but use of nonhuman primates for studies of variola have priority.	Moderately useful (especially vaccinia) for obtaining information about antiviral activity of candidate drugs and vaccines.[b]
Monkeypox	Limited usefulness: some potential to help in identifying useful interventions against smallpox, but other approaches are more important.[c]	Moderately useful for obtaining information about antiviral activity of candidate drugs and vaccines.	Most useful in suggesting likely benefits from candidate therapeutics and vaccines against variola in the human population.[d]
Variola	Not an available option for developing therapeutics or vaccines.	Most useful in suggesting likely benefits from candidate therapeutics and vaccines against variola in the human population.	Not an available option for developing therapeutics or vaccines.

[a]Dual infection of nonhuman primates with simian immunodeficiency virus (SIV) and vaccinia produces disease with features that mimic human smallpox.

[b]Disseminated vaccinia infection in humans produces disease with some features reminiscent of smallpox.

[c]Monkeypox virus infection of ground squirrels has been used to assess monkeypox vaccines.

[d]This cell, monkeypox in humans, refers to naturally occurring disease. The study of this disease might provide an opportunity to assess diagnostics, therapeutics, and vaccines for their utility in both monkeypox and variola.

recombinations of the parent virus genomes and have not been characterized by sequencing. Furthermore, while studies with these recombinants might contribute to understanding of variola pathogenesis, such investigations could not substitute for those using variola virus and would have to be carried out in BSL-4 facilities that should be used instead to support experiments with variola virus that are essential for developing smallpox therapeutics and vaccines.

USEFULNESS OF VARIOUS MODELS

Table 4-1 summarizes the usefulness of animal models and human infection with monkeypox or vaccinia for understanding the pathogenesis of variola and other orthopoxviruses and for developing therapeutics and vaccines. Although some of these approaches are more useful than other, none is ideal in recreating the equivalent of human smallpox.

REFERENCES

Abdalrhman, I., I. Gurt, and E. Katz. 2006. Protection induced in mice against a lethal orthopox virus by the Lister strain of vaccinia virus and modified vaccinia virus Ankara (MVA). *Vaccine* 24(19):4152–4160.

Adams, M. M., A. D. Rice, and R. W. Moyer. 2007. Rabbitpox virus and vaccinia virus infection of rabbits as a model for human smallpox. *Journal of Virology* 81(20):11084–11095.

Bedson, H. S., and K.R. Dumbell. 1964a. Hybrids Derived from the Viruses of Alastrim and Rabbit Pox. *Journal of Hygiene (London)* 62:141–146.

Bedson, H. S., and K. R. Dumbell. 1964b. Hybrids derived from the viruses of variola major and cowpox. *Journal of Hygiene (London)* 62:147–158.

Belyakov, I. M., P. Earl, A. Dzutsev, V. A. Kuznetsov, M. Lemon, L. S. Wyatt, J. T. Snyder, J. D. Ahlers, G. Franchini, B. Moss, and J. A. Berzofsky. 2003. Shared modes of protection against poxvirus infection by attenuated and conventional smallpox vaccine viruses. *Proceedings of the National Academy of Sciences of the United States of America* 100(16):9458–9463.

Bray, M., M. Martinez, D. F. Smee, D. Kefauver, E. Thompson, and J. W. Huggins. 2000. Cidofovir protects mice against lethal aerosol or intranasal cowpox virus challenge. *Journal of Infectious Diseases* 181:10–19.

Earl, P. L., J. L. Americo, L. S. Wyatt, L. A. Eller, J. C. Whitbeck, G. H. Cohen, R. J. Eisenberg, C. J. Hartmann, D. L. Jackson, D. A. Kulesh, M. J. Martinez, D. M. Miller, E. M. Mucker, J. D. Shamblin, S. H. Zwiers, J. W. Huggins, P. B. Jahrling, and B. Moss. 2004. Immunogenicity of a highly attenuated MVA smallpox vaccine and protection against monkeypox. *Nature* 428:182–185.

Fenner, F., 1949. Mouse-pox (infectious ectromelia of mice): A review. *The Journal of Immunology* 63: 341–373.

Ferrier-Rembert, A., R. Drillien, B. Meignier, D. Garin, and J.M. Crance. 2007. Safety, immunogenicity and protective efficacy in mice of a new cell-cultured Lister smallpox vaccine candidate. *Vaccine* 25(49):8290–8297.

Hahon, N. 1961. Smallpox and related poxvirus infections in the simian host. *Bacteriological Reviews* 25:459–476.

Hahon, N., and B. J. Wilson. 1960. Pathogenesis of variola in *Macaca irus* monkeys. *American Journal of Hygiene* 71:69–80.

Hooper, J. W., D. M. Custer, and E. Thompson. 2003. Four-gene-combination DNA vaccine protects mice against a lethal vaccinia virus challenge and elicits appropriate antibody responses in nonhuman primates. *Virology* 306:181–195.

Hutin, Y. J., R. J. Williams, P. Malfait, R. Pebody, V. N. Loparev, S. L. Ropp, M. Rodriguez, J. C. Knight, K. Tshioko, A. S. Khan, M. V. Szczeniowski, and J. J. Esposito. 2001. Outbreak of human monkeypox, Democratic Republic of Congo, 1996 to 1997. *Emerging Infectious Diseases* 7:434–438.

Hutson, C. L., Lee, K. N., J. Abel, D. S. Carroll, J. M. Montgomery, V. A. Olson, Y. Li, W. Davidson, C. Hughes, M. Dillon, P. Spurlock, J. J. Kazmierczak, C. Austin, L. Miser, F. E. Sorhage, J. Howell, J. P. Davis, M. G. Reynolds, Z. Braden, K. L. Karem, I. K. Damon, and R. L. Regnery. 2007. Monkeypox zoonotic associations: Insights from laboratory evaluation of animals associated with the multi-state U.S. outbreak. *American Journal of Tropical Medicine and Hygiene* 76:757–768.

Jahrling, P. B., L. E. Hensley, M. J. Martinez, J. W. LeDuc, K. H. Rubins, D. A. Relman, and J. W. Huggins. 2004. Variola virus infection of cynomolgus macaques: A model for human smallpox. *Proceedings of the National Academy of Sciences of the United States of America* 101:15196–15200.

Jones, T. C., R. D. Hunt, and N. W. King (editors). 1997. *Veterinary pathology*. Baltimore, MD: Williams and Wilkins. p. 204.

Lancaster, M. C., E. A. Boulter, J. C. Westwood, and J. Randles. 1966. Experimental respiratory infection with poxviruses. II. Pathological studies. *British Journal of Experimental Pathology* 47(5):466–471.

Law, M., M. M. Putz, and G. L. Smith. 2005. An investigation of the therapeutic value of vaccinia-immune IgG in a mouse pneumonia model. *Journal of General Virology* 86:991–1000.

Levine, R. S., A. T. Peterson, K. L. Yorita, D. Carroll, I. K. Damon, and M. G. Reynolds. 2007. Ecological niche and geographic distribution of human monkeypox in Africa. *PLoS One* 2:e176.

McCurdy, L. H., Rutigliano, J. A., T. R. Johnson, M. Chen, and B. S. Graham. 2004. Modified vaccinia virus Ankara immunization protects against lethal challenge with recombinant vaccinia virus expressing murine interleukin-4. *Journal of Virology* 78:12471–12479.

Parker, S., A. Nuara, R. M. Buller, and D. A. Schultz. 2007. Human monkeypox: An emerging zoonotic disease. *Future Microbiology* 2:17–34.

Phelps, A., A. J. Gates, M. Hillier, L. Eastaugh, and D. O. Ulaeto. 2005. Comparative efficacy of replicating smallpox vaccine strains in a Murine Challenge Model. *Vaccine* 23(27):3500–3507.

Phelps, A. L., A. J. Gates, M. Hillier, L. Eastaugh, and D. O. Ulaeto. 2007. Comparative efficacy of modified vaccinia Ankara (MVA) as a potential replacement smallpox vaccine. *Vaccine* 25(1):34–42.

Rubins, K., L. E. Hensley, P. B. Jahrling, A. A. Whitney, T. W. Geisbert, J. W. Huggins, A. Owen, J. W. LeDuc, P. O. Brown, and D. A. Relman. 2004. The host response to smallpox: Analysis of the gene expression program in peripheral blood cells in a monkey model. *Proceedings of the National Academy of Sciences of the United States of America* 101:15190–15195.

Smee, D. F., K. W. Bailey, M. H. Wong, and R. W. Sidwell. 2001. Effects of cidofovir on the pathogenesis of a lethal vaccinia virus respiratory infection in mice. *Antiviral Research* 52:55–62.

Stanford, M. M., S. J. Werden, and G. McFadden. 2007. Myxoma virus in the European rabbit: Interactions between the virus and its susceptible host. *Veterinary Research* 38:299–318.

Stittelaar, K. J., G. van Amerongen, I. Kondova, T. Kuiken, R. F. van Lavieren, F. H. Pistoor, H. G. Niesters, G. van Doornum, B. A. van der Zeijst, L. Mateo, P. J. Chaplin, and A. D. Osterhaus. 2005. Modified vaccinia virus Ankara protects macaques against respiratory challenge with monkeypox virus. *Journal of Virology* 79:7845–7851.

Tesh, R. B., D. M. Watts, E. Sbrana, M. Siirin, V. L. Popov, and S. Y. Xiao. 2004. Experimental infection of ground squirrels (Spermophilus tridecemlineatus) with monkeypox virus. *Emerging Infectious Diseases* 10:1563–1567.

Westwood, J. C., E. A. Boulter, E. T. Bowen, and H. B. Maber. 1966. Experimental respiratory infection with poxviruses. I. Clinical virological and epidemiological studies. *British Journal of Experimental Pathology* 47(5):453–465.

Wyatt, L. S., P. L. Earl, L. A. Eller, and B. Moss. 2004. Highly attenuated smallpox vaccine protects mice with and without immune deficiencies against pathogenic vaccinia virus challenge. *Proceedings of the National Academy of Sciences of the United States of America* 101:4590–4595.

5

Genomic Analysis

Understanding the biology of variola virus and its genetic relatives is critical for developing countermeasures for smallpox, and genome sequencing is one of the most effective initial steps in achieving an understanding of the biology of any life form. The genome sequence is the "blueprint" that describes the entire suite of biological capabilities of any cellular organism or virus. With the recent rapid growth in genome sequencing capabilities, it has become increasingly clear that genomics (the study of genome sequences and of the functions they encode) can provide unexpected insights into host–pathogen interactions, the evolutionary history of a virus, and evolutionary relationships among viruses. Poxvirus genome sequence analysis can be used to identify potential targets for the development of therapeutics and vaccines, and offers the promise of revealing the molecular events underlying smallpox infection and pathogenesis and the host response.

The last 10 years have seen enormous technological advances leading to much more rapid and much less expensive methods for genome sequencing. Today it is possible to sequence the complete genomes of all variola strains in both authorized repositories in less time and at a fraction of the cost required to sequence one variola virus genome in 1999. While Chapter 3 provided an overview of poxvirus genomics, this chapter presents a more detailed review of variola genomics and the progress that has been made in this area since 1999. Since then, a more refined understanding of the evolution and viral population structure of variola virus has emerged from genomic analysis. In addition, as with monkeypox virus strain analysis (Chapter 3), the sequencing of geographically distinct isolates of variola has yielded some important clues as to the viral determinants of human disease. This progress and these discoveries, as well as the remaining unanswered questions about the links between viral genetics and disease manifestations, serve as the foundation for the committee's conclusion and recommendation regarding variola genomics (see Chapter 10). The conclusion of the 1999 IOM committee on this subject provides context for the discussion that follows:

> **Genomic sequencing and limited study of variola surface proteins derived from geographically dispersed specimens is an essential foundation for important future work. Such research could be carried out now, and could require a delay in the destruction of known stocks, but would not necessitate their indefinite retention.**

The 1999 committee believed that the sequencing of multiple strains would provide greater understanding of genetic variation and variation in genome structure and content among strains, particularly at the terminal ends (where genes associated with pathogenicity and virulence often reside).

This chapter reviews work done to date to analyze the genome of variola virus, the additional work on variola genomics needed to support the development of smallpox countermeasures and increase understanding of smallpox infection and pathogenesis and the host response, and the need for live variola virus in this work.

SEQUENCE ANALYSIS

The genome of variola virus (VARV) contains approximately 186 kilobases of double-stranded DNA and approximately 200 nonoverlapping open reading frames (ORFs). Each end of the genome is covalently closed, and regions of inverted terminal repeats (ITRs) flank the central coding region. VARV is unique among the poxviruses in that its ITRs do not contain ORFs (Massung et al., 1996; Esposito et al., 2006).

Work published on the genomics of variola virus since 1999 has been restricted largely to isolates held in the CDC repository. In work published in 2006, full genome sequences were determined for 43 geographically distinct VARV isolates held in the CDC repository. Several other variola genome sequences are also available at present, for a total of at least 49. These strains are not necessarily representative of the extant global variola virus population from the last half of the twentieth century, but were selected because they were isolated from cases of smallpox that occurred in geographically diverse regions of the world for which reasonably reliable epidemiological data and case fatality rates were available.

The analysis of these 43 genome sequences, in combination with the two previously determined full genome sequences, revealed a high degree of conservation of centrally located coding region sequences (mid-CRS) among strains, supporting a role for these ORFs in ensuring the fitness of the virus through its life cycle (Esposito et al., 2006). Furthermore, the terminal CRS regions, adjacent to the ITRs, display variation among isolates with differing case fatality rates. Nearly 90 percent of VARV predicted ORFs can be identified in the genomes of other orthopoxviruses, with the remaining VARV ORFs being found as partial forms in other orthopoxvirus genomes (Esposito et al., 2006). Over the entire roughly 186,000 base genome, pairs of variola strains differ by as many as about 700 single nucleotide polymorphisms (SNPs) and about 90 insertions or deletions (indels), and by as few as a handful of each (overall, among all viruses, there are 1,782 specific SNPs and 4,812 specific indels). Taken together, these data indicate restricted variability in the overall genome sequence and support the notion that the terminal CRS regions contain ORFs important for host interaction and pathogenesis, while the mid-CRS-region ORFs are critical for expressing conserved proteins important to virus replication (Esposito et al., 2006; Moss, 2007).

Phylogenetic analysis of these epidemiologically distinct VARV isolates, isolated from patients over a period of 30 years, reveals two primary clades of VARV with distinct clustering based on the geographic region from which the source patients derived. One clade ("A" in Figure 5-1) includes variola major virus isolates from Asia that were associated with clinically severe (high case fatality rate) cases of smallpox and variola isolates from east, central, and southern Africa associated with disease of variable severity. The other clade consists of two subclades, one comprising alastrim minor isolates from South America ("B" in Figure 5-1), which were associated with mild smallpox disease, and the other ("C" in Figure 5-1) comprising variola isolates from west Africa that were associated with intermediate disease severity (Li et al., 2007). The tendency of strains to cluster based on geography has provided clues as to how the virus spread among humans around the world. Some relationships between variola genome sequences and disease severity have been identified, although these associations are only broadly defined at present because the range of case fatality rates within some clades is large. For example, the case fatality rates of isolates from Asian clade C range from <1 percent to 38 percent, compared with 8 percent to 12 percent for west African clade A and 0.8 percent for South American clade B. Furthermore, both viral and host features, such as age and nutritional status, are important in determining clinical outcome.

Based on genome sequence comparisons, variola virus is most closely related to camelpox and taterapox viruses, with which it shares approxi-

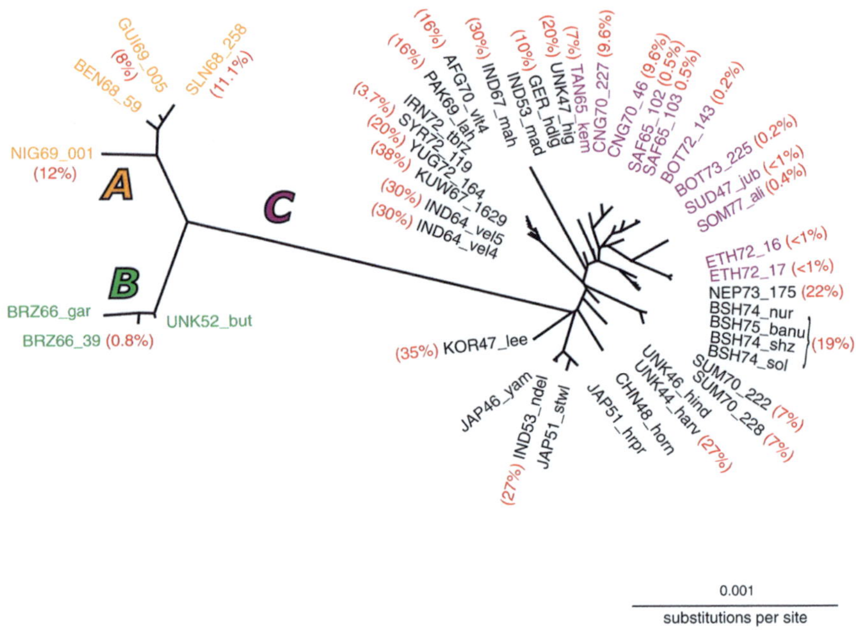

FIGURE 5-1 Variola virus evolutionary relationships based on an alignment of a conserved mid-region of genomic DNA sequences of 45 isolates (from Esposito et al., 2006; Figure 3). The isolates are from smallpox case-patients in west Africa (clade A), South America (clade B), and Asia (clade C); the Asian clade C includes a subgroup of non–west African African variants. Case fatality rates associated with some isolates are indicated in parentheses.
SOURCE: Esposito et al., 2006. Reprinted with permission from AAAS.

mately 98 percent overall sequence similarity (Esposito et al., 2006). VARV probably arose from an ancestral rodent-associated variola-like virus in Africa between 16,000 and 68,000 years ago (Li et al., 2007). The data also indicate a different evolutionary history for the mid-CRS and terminal CRS regions of the VARV genome among the different clades (Esposito et al., 2006; Li et al., 2007).

According to WHO (Lavanchy, 2008; WHO, 2008), 891 isolates (120 strains) were held at the authorized repository at VECTOR until 2008, when it was reported that 200 nonviable and duplicate samples were destroyed. It is not clear whether or to what degree the remaining 691 isolates (120 strains) at VECTOR and the 406 unsequenced isolates (184 unsequenced strains) at CDC offer novel features not found among the at least 48 isolates sequenced thus far (2 sequenced prior to 2000 and

47 since then; 1 sequenced twice [Bangladesh, 1975]). However, it is reasonable to assume that additional diversity would be discovered if these strains were sequenced.

In comparison with the significant effort required to determine a complete poxvirus genome sequence in 1993, the effort necessary today to sequence all remaining variola isolates would be relatively minor. In fact, if DNA were made available, current next-generation sequencing technology would enable the determination of complete genome sequences for all remaining variola strains in a total of several weeks by one laboratory, at a low cost. Furthermore, it should be feasible to obtain sufficient DNA from each strain using current whole-genome amplification techniques, obviating the need for in vitro cultivation of these strains for the purpose of genome sequencing. Since fewer African than Asian isolates have been sequenced, obtaining more genome sequence information about these isolates should be a priority. Given these advances, the scientific benefits of sequencing all remaining isolates today vastly outweigh the costs.

BEYOND GENOMIC ANALYSIS

Significant progress has been made in poxvirus genomics. More than 111 poxvirus genome sequences are now available (http://www.poxvirus.org/), and 48 VARV genomes are available for public access in GenBank (http://www.ncbi.nlm.nih.gov/Genbank/). Further, plaque phenotyping and assessment of comet morphology for the sequenced VARV isolates have been completed, deepening understanding of the relationship between the biological properties of VARV isolates and their genome sequences (Olson et al., 2009). Analyzing VARV genome sequences yields insights into virulence; greatly improves the reliability of nucleic acid-based detection and diagnostic assays (see Chapter 8); and makes it possible to begin to understand better the biology of this virus, facilitating the development of new therapeutics (see Chapter 6).

However, meaningful exploitation of genome sequence requires an assessment of functional attributes. Additional work is necessary to understand the molecular mechanisms of viral pathogenesis and replication in order to support the development of effective countermeasures and means of detection. Among the possible approaches are targeted assessments of genes or gene products of interest, as well as genome-wide "functional genomics" methodologies, such as analysis of (virus and host) genome-wide transcript and small RNA abundance, profiling of proteins and phosphoproteins, and analysis of protein–DNA binding patterns. Recent advances in computational methods allow identification of gene networks, metabolic pathways, and genetic modules and nodes (Litvin et al., 2009), all of which may reveal novel, critical targets for therapeutic intervention in both virus and host.

Additional insights into VARV pathogenesis will come from characterization of the VARV proteome, the entirety of the proteins expressed by VARV. Work of this sort has been undertaken with vaccinia virus and other orthopoxvirus proteins (Chung et al., 2006; Resch et al., 2007). However, the ability to extrapolate findings from vaccinia virus to variola virus is unclear, but almost certainly limited.

NEED FOR LIVE VARIOLA VIRUS

In the past, genome projects have required large amounts of high-quality genomic DNA, which in turn has usually necessitated propagation of the agent to high titer in the laboratory. Today, however, the ability to isolate and amplify DNA from microbes is greatly improved, such that adequate DNA can be generated from a single bacterial cell for full-genome projects (Marcy et al., 2007). In general, live variola virus is not needed for variola genome sequence analysis as long as DNA of adequate quantity and quality is available. The latter need can be met either by cultivation and DNA harvesting or by DNA amplification methods, the products of which can be saved in the form of genomic clones or amplified DNA. On the other hand, live variola virus would be needed to perform functional studies (such as studying RNA or protein expression or host interactions) for the purpose of understanding pathogenesis so as to identify new targets for therapeutics.

REFERENCES

Chung, C.S., C. H. Chen, M. Y. Ho, C. Y. Huang, C. L. Liao, and W. Chang. 2006. Vaccinia virus proteome: Identification of proteins in vaccinia virus intracellular mature virion particles. *Journal of Virology* 80(5):2127–2140.

Esposito, J. J., S. A. Sammons, A. M. Frace, J. D. Osborne, M. Olsen-Rasmussen, M. Zhang, D. Govil, I. K. Damon, R. Kline, M. Laker, Y. Li, G. L. Smith, H. Meyer, J. W. Leduc, and R. M. Wohlhueter. 2006. Genome sequence diversity and clues to the evolution of variola (smallpox) virus. *Science* 313:807–812.

Lavanchy, D. 2008. WHO Smallpox program: Live variola virus. Presentation to the Committee, December 18–19, 2008.

Li, Y., D. S. Carroll, S. N. Gardner, M. C. Walsh, E. A. Vitalis, and I. K. Damon. 2007. On the origin of smallpox: Correlating variola phylogenics with historical smallpox records. *Proceedings of the National Academy of Sciences of the United States of America* 104:15787–15792.

Litvin, O., H. C. Causton, B. J. Chen, and D. Pe'er. 2009. Modularity and interactions in the genetics of gene expression. *Proceedings of the National Academy of Sciences of the United States of America*. [Epub ahead of print].

Marcy, Y., C. Ouverney, E. M. Bik, T. Lösekann, N. Ivanova, H. G. Martin, E. Szeto, D. Platt, P. Hugenholtz, D. A. Relman, and S. R. Quake. 2007. Dissecting biological "dark matter" with single-cell genetic analysis of rare and uncultivated TM7 microbes from the human mouth. *Proceedings of the National Academy of Sciences of the United States of America* 104:11889–11894.

Massung, R. F., V. N. Loparev, J. C. Knight, A. V. Totmenin, V. E. Chizhikov, J. M. Parsons, P. F. Safronov, V. V. Gutorov, S. N. Shchelkunov, and J. J. Esposito. 1996. Terminal refion sequence variations in variola virus DNA. *Virology* 22 (2):291–300.

Moss, B. M. 2007. Poxviridae: The viruses and their replication. In *Fields' virology*, 5th edition, edited by B. N. Fields, D. M. Knipe, P. M. Howley, and D. E. Griffin. Philadelphia: Lippincott Williams & Wilkins. Pp. 2906–2946.

Olson, V.A., K. L. Karem, S. K. Smith, C. M. Hughes, and I. K. Damon. 2009. Smallpox virus plaque phenotypes: genetic, geographical and case fatality relationships. *Journal of General Virology* 90(Pt 4):792–798.

Resch, W., K. K. Hixson, R. J. Moore, M. S. Lipton, and B. Moss. 2007. Protein composition of the vaccinia virus mature virion. *Virology* 358(1):233–247.

WHO (World Health Organization). 2008. WHO Advisory Committee on Variola Virus Research Report of the Tenth Meeting. November 19–20. Geneva, Switzerland: WHO. http://www.who.int/csr/resources/publications/WHO_HSE_EPR_2008_9/en/index.html (accessed March 23, 2009).

6

Development of Therapeutics

Historical evidence suggests that, in response to an accidental or intentional release of variola virus, smallpox vaccination would be an effective public health measure to protect at-risk populations. To be protective, however, vaccination must occur within 4 days of exposure to the virus (Fenner et al., 1998; Mortimer, 2003). In addition, there are contraindications to the administration of current smallpox vaccines, particularly among immunocompromised individuals. These individuals would need alternative protective measures following exposure to variola virus.

To reduce the significant morbidity and mortality in cases of smallpox, safe and effective therapeutic agents are required. By accelerating clearance of the virus from ill individuals, such agents may also limit infectivity and transmission of disease. Antiviral agents can also be useful for prophylaxis after exposure has occurred. The availability of these agents has the potential to be important for both the treatment and prophylaxis of smallpox in exposed persons identified after the 4-day period when vaccination is effective, and could be a valuable component of any effective control strategy. In the last decade, substantial progress has been made in the development of therapeutics with the potential to meet this need (see Tables 6-1 and 6-2). However, these efforts have yet to yield an FDA-licensed agent for the treatment or prevention of smallpox and other orthopoxviruses.

The 1999 IOM report identified the development of antiviral agents as the most significant reason to retain stocks of live variola virus, primarily because of the lack of availability of an effective therapeutic agent (either currently or historically) that could serve as a standard for purposes of comparison:

The most compelling reason for long-term retention of live variola virus stocks is their essential role in the identification and development of antiviral agents for use in anticipation of a large outbreak of smallpox. It must be emphasized that if the search for antiviral agents with activity against live variola virus were to be continued, additional public resources would be needed.

The 1999 report also suggested that having more than one antiviral agent would be desirable because of the potential for the emergence of drug-resistant variola strains. Replication-deficient forms of variola virus could be used to develop new agents; ultimately, however, the live intact virus would be required to ensure confidence in the results. The 1999 report also noted that, given the lack of incentive for the development of smallpox therapeutics in the private sector, significant public resources would need to be mobilized.

This chapter reviews potential therapeutics for smallpox, regulatory requirements for the development of such therapeutics, and the need for live variola virus in this work.

POTENTIAL THERAPEUTICS FOR SMALLPOX

Potential therapeutics for smallpox include two drugs approved by the FDA for other purposes, newly developed drugs, agents to block newly identified poxvirus targets, and drugs that enhance or modulate the host's immune response.

Use of Drugs Approved by the FDA for Other Purposes

Because de novo drug development is an expensive and time-consuming process (costing in excess of $500 million and requiring approximately 8–10 years of continuous effort) (Henderson and Fenner, 2001), the use of licensed drugs approved for other purposes represents an attractive option for antivirals against variola.

Cidofovir is a DNA polymerase inhibitor, licensed for the treatment of cytomegalovirus-induced retinitis in HIV-infected individuals (Tesh et al., 2004). Cidofovir also exhibits in vitro antiviral activity against poxviruses, and is effective against cowpox and vaccinia virus infections in mice (LeDuc et al., 2002; Baker et al., 2003; Quenelle et al., 2003; Magee et al., 2005). Under an Investigational New Drug (IND) protocol from the FDA, cidofovir can be used to treat acute smallpox and complications arising from vaccinia infection when a patient has not responded to administration of vaccinia immune globulin (VIG) (LeDuc et al., 2002; reviewed in Sliva and Schnierle, 2007). However, the utility of cidofovir for treating smallpox is complicated by the fact that the drug is available only in a topical or intravenous formulation. A topical formulation would have no role in treating a systemic disease such as smallpox. Intravenous cidofovir must be given as a 1-hour infusion in combination with multiple doses of probenecid and requires sustained intravenous hydration and monitoring of renal function. Even when given intravenously, the drug does not cross the blood–brain barrier. Although cidofovir's long half-life has the advantage of allowing weekly dosing, problems with administration and toxicity make large-scale use of this agent difficult. It is not likely to be usable in resource-poor settings. The emergence of resistance is also a concern because exposure of vaccinia to cidofovir resulted in the emergence of mutations in the DNA polymerase gene, which is the target of the drug (Becker et al., 2008).

Gleevec (also referred to as STI-571 or imatinib mesylate) is an FDA-approved treatment for chronic myeloid leukemia that exhibits antiviral activity against poxviruses. Gleevec blocks the action of Abl-family tyrosine kinases (Druker et al., 1996) and thus blocks the egress of vaccinia virus from infected cells in vitro (McFadden, 2005; Reeves et al., 2005; Yang et al., 2005). It has also undergone in vitro testing against the monkeypox and variola viruses with similar effects (Reeves et al., 2006). In addition, Gleevec treatment promoted survival of mice following intranasal challenge with vaccinia virus, and it has been suggested as a potential therapeutic for postvaccination complications associated with vaccinia (Reeves et al., 2005). The drug does not appear to interfere with the development of immunity that protects against subsequent challenge. However, the protective benefit of Gleevec was evident only at lower virus titers and only when the drug was given less than 48 hours after exposure. Studies of Gleevec in rabbits infected with rabbitpox and in mice infected with ectromelia showed much lower antiviral activity than in other animal models (personal communication, Dr. Daniel Kalman, Emory University, February 2009). The reduced activity against higher titers of the inoculum virus, the requirement for administration shortly after inoculation, and the variable protection in poxvirus models raise concerns about Gleevec's potential for treating smallpox.

Newly Developed Therapeutics

To overcome the challenges associated with cidofovir discussed above, orally bioavailable cidofovir derivatives have recently been developed (HDP-cidofovir/CMX-001) (Ciesla et al., 2003; Buller et al., 2004; Kern et al., 2004). CMX-001 also displays enhanced antiviral activity against variola virus in comparison with cidofovir (Bradbury, 2002; Morris, 2002; Sliva and Schnierle, 2007). The inhibitory activity of hexadecyloxypropyl-CDV is 40–100 times grater than that of CDV in vitro in cells infected with

TABLE 6-1 Development of Therapeutics for Smallpox

Drug	Chemical Structure	Mode of Action	Reference
Cidofovir[a] (CDV)	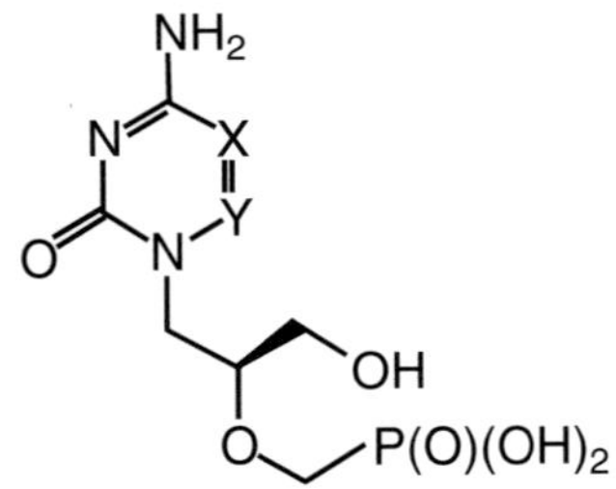 **1**, X = N, Y = CH **2**, X = CH, Y = N	ATP analog that inhibits DNA polymerase. When incorporated into the template strand, blocks DNA elongation and 3′–5′ proof-reading exonuclease activity. Drug resistance seen in the VV E9L gene (DNAP). FDA approved for cytomegalovirus retinitis in persons with HIV.	Baker et al., 2003; Magee et al., 2005, 2008; Krecmerová et al., 2007a
CDV derivatives[b,c]: CMX001 and HPMP-5-azaCb	CMX-001 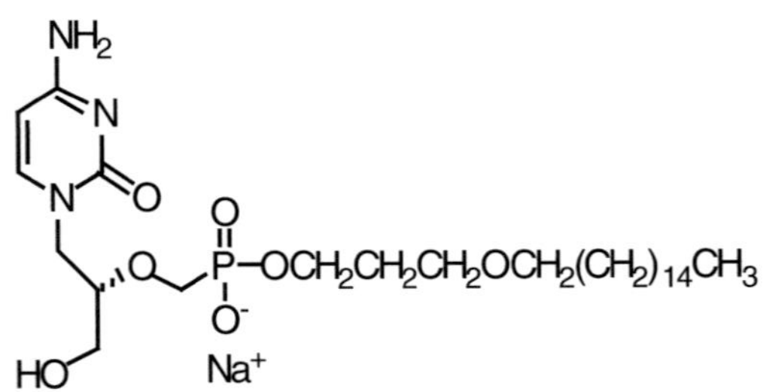	Same target as CDV. Esterification makes these derivatives more lipophilic and increases uptake roughly 50-fold. CMX001 is 100-fold more active than CDV and does not produce renal toxicity.	Kern et al., 2002; Kern, 2003; Krecmerová et al., 2007b; Quenelle et al., 2007b; Magee et al., 2008; Naesens et al., 2008; Parker et al., 2008

	HPMP-5-azaCb 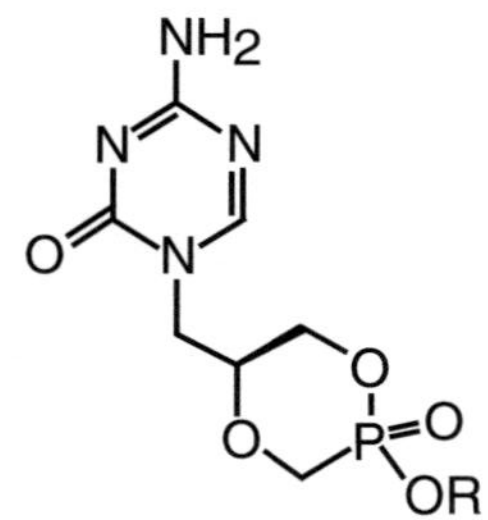		
STI-571 (Gleevec or imatinib mesylate)	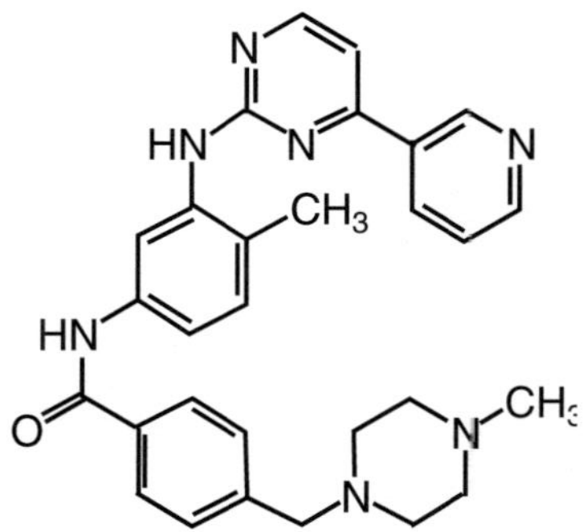	Blocks the Abl-family tyrosine kinases needed for the actin motility of intracellular viral particles (IMV), thus blocking egress of IMV from cells. FDA approved for chronic myeloid leukemia.	McFadden, 2005; Reeves et al., 2005

continued

TABLE 6-1 Continued

Drug	Chemical Structure	Mode of Action	Reference
ST-246[d]	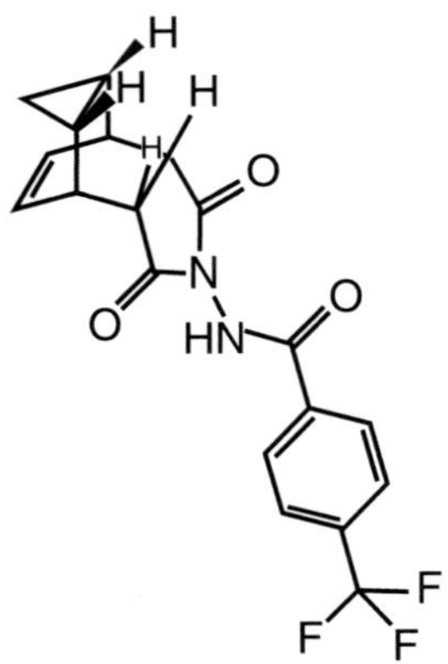	Inhibits virus release by targeting a pox protein (p37 or 60L for cowpox or F13L for vaccinia) that is essential for envelopment of IMV.	Yang et al., 2005; Quenelle et al., 2007a,b

4′-thioIDU[e]	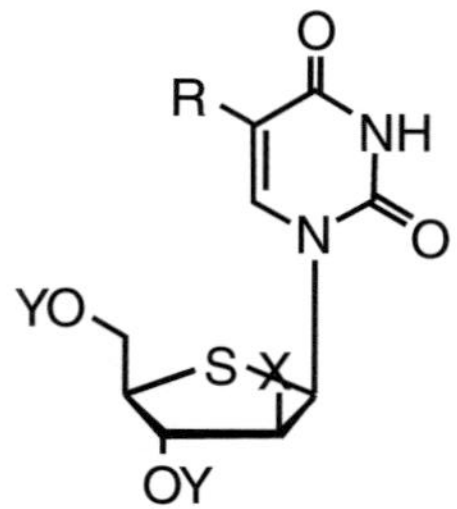	Targets the Thymidine Kinase (TK) gene to inhibit DNA synthesis.	Kern et al., 2009

[a]Cidofovir {HPMPC, CDV, 1-(S)-[3-hydroxy-2-(phosphonomethoxy) propyl] cytosine}.
[b]{1-(S)-[3-hydroxy-2-(phosphonomethoxy)propyl]-5-azacytosine; HPMP-5-azaC} is an analog of CDV.
[c]CMX001 is hexadecyloxypropyl CDV and has much better oral bioavailability than CDV.
[d]ST-246 [N-(3,3a,4,4a,5,5a,6,6a-octahydro-1,3-di oxo-4,6-etheneocycloprop[f]isoindol-2(1H)-yl)-benzamide] used at 20 μg/ml in cell cultures reduces EMV by 6 logs and reduces IMV by 2 logs.
[e]1-(2′-deoxy-4′-thio-β-D-ribofuranosyl)-5-iodouracil (4′-thioIDU). The R can be F, Br, I, CH3, CF3, or a phenyl group. The X can be H or OH. The Y can be H or an acetyl group.

TABLE 6-2 Clinical Aspects of Therapeutic Agents for Smallpox in Humans

Drug	Dose and Route of Administration	Issues
Cidofovir	5 mg/kg given intravenously weekly/biweekly; also used topically and intralesionally	Severe renal toxicity; co-administration of probenecid necessary; hydration requirement; used off label for recurrent laryngeal papillomatosis, molluscum contagiosum, and human papillomavirus
Esters of Cidofovir (CMX001)	Dose not yet defined; taken orally	Phase I and II human clinical trials under way; no reported renal toxicity to date; higher bioavailability than CDV
Gleevec	400–800 mg/day; taken orally	Chemotherapeutic agent; side effects include edema, cytopenia, and hepatotoxicity
ST-246	500–2000 mg/day; taken orally for 14 days; other routes of administration (intravenous, liquid suspension) being considered	Minimal toxicity seen in human dosing trials

variola, cowpox, vaccinia, or ectromelia virus. Protection of mice from lethal mousepox infection has been demonstrated (Parker et al., 2008), and CMX-001 was effective against mousepox in the C57BL/6 strain, which is considered to have a course of infection more similar to that of variola than its progression in other mousepox strains when given 4 days after inoculation (Parker et al., 2009). Other derivatives of cidofovir could prove effective as well (Lebeau et al., 2006; Stittelaar et al., 2006; Hostetler et al., 2007; Hostetler, 2009). Orally bioavailable cidofovir derivatives have shown negligible renal toxicity, a significant advantage over the intravenous formulation. CMX-001 has been given to a patient with eczema vaccinatum who did not respond to ST-246 (CDC, 2009). A recently completed human volunteer phase I multidose study with more than 100 subjects demonstrated no significant adverse events, and phase II trials are being initiated (Painter and Hostetler, 2004; Ruiz et al., 2007).

ST-246, which was discovered from a high-throughput screen of 356,240 small-molecule inhibitors of vaccinia virus replication, is currently being used in human trials. This antiviral drug targets the vaccinia virus protein F13, which is essential for envelopment and egress of the intracellular mature virions (MV) and subsequent viral spread (Yang et al., 2005). Cell cultures infected with six different variola isolates or seven different monkeypox isolates showed reduced cytopathic effects, virus production, and comet tail formation after treatment with nanomolar amounts of ST-246. ST-246 is 8,000 times more potent than cidofovir in vitro against poxviruses, is orally bioavailable, and is stable at room temperature. It has proven to be effective in blocking replication of all orthopoxviruses that have been tested in vitro (Duraffour et al., 2007) and in protecting mice (Yang et al., 2005; Quenelle et al., 2007a), rabbits (Nalca et al., 2008), and ground squirrels (Sbrana et al., 2007) from orthopoxvirus challenge. Animals infected with monkeypox, cowpox, ectromelia, and variola viruses that received ST-246 were protected from lethal infection and also mounted a protective immune response (Bolken and Hruby, 2008; Nalca et al., 2008). ST-246 in combination with CMX-001 displays synergistic antiviral effects against vaccinia and cowpox in animals without increasing toxicity (Quenelle et al., 2007b; Whitley, 2008). In 2007, a 14-day course of ST-246 was used in conjunction with cidofovir and VIG under an emergency IND to treat a severe case of eczema vaccinatum in an infant who was infected with vaccinia as a result of contact transmission (Vora et al., 2008). Since cidofovir and VIG were co-administered with ST-246, however, it is not clear that the resolution of the infection is attributable entirely or even partially to ST-246. Human phase I trials of ST-246 have been completed. The drug was given to 31 healthy individuals in a single dose ranging from 500 mg to 2000 mg daily in a fasting and nonfasting state, with an 8-person placebo group used for comparison (Jordan et al., 2008). Side effects were minimal, and only reversible neutropenia was seen more often in the treated than in the placebo group.

Important information on ST-246 has been obtained: the variola gene product targeted by ST-246 is known, and the doses have been shown to be effective against poxviruses in mice and nonhuman primates. However, clinical data are needed on the use of ST-246 in humans; studies to provide these data are under development for naturally occurring human monkeypox but will be difficult to implement and monitor. An important caveat for antiviral drugs such as ST-246 that exhibit high potency in vitro is that they can be tested only in model systems or against other poxvirus infections in humans, and it is impossible to know with certainty how they would perform against smallpox in the event of its reemergence.

Work on the development of new drugs has also continued in Russia. VECTOR reports having conducted screening of more than 5,000 chemical compounds for their antiviral activity, and about 80 compounds active against surrogate orthopoxviruses (vaccinia virus, cowpox virus, and ectromelia virus) are said to have been identified. In testing done in cell culture, VECTOR reports that 60 compounds demonstrated antiviral activity against variola virus (Zakirova et al., 2004; Ivanov et al., 2005, 2008; personal communication, Ilya Drozdov, WHOCC for Orthopoxvirus Diagnosis and Repository for Variola Virus Strains and DNA, March 27, 2009).

Agents to Block Newly Identified Poxvirus Targets

Further research on poxvirus replication has made it possible to identify possible poxvirus drug targets, and assessment of molecules blocking these targets has begun (Yang et al., 2005; Sliva and Schnierle, 2007; Tse-Dinh, 2008). Three enzymes involved in vaccinia virus replication have been identified and crystallized: thymidine kinase (TK), deoxyuridine triphosphatase (DUTPase), and uracil DNA glycosylase (UDG) (Whitley, 2008). DNA polymerase nucleoside inhibitors (Fan et al., 2006; Prichard et al., 2007), nucleoside inhibitors of S-adenosyl-L-homocysteine hydrolase (De Clerq and Holy, 2005; Roy et al., 2005; Yang and Schneller, 2005; Arumugham et al., 2006), targets of topoisomerase I (Da Fonseca and Moss, 2003; Bond et al., 2006; Fujimoto et al., 2006; Perry et al., 2006), and other egress inhibitors (Bailey et al., 2007) have been evaluated for their ability to block poxvirus replication. It has been suggested that the new 4′ thioIDU (TK inhibitor) might be an additional component of combination therapy since it can block replication of CMX–001- and ST-246-resistant mutants (Kern et al., 2009). The use of combination antiviral therapy is favored as it may slow the development of drug-resistant strains of variola and other orthopoxviruses. When administered intraperitoneally or orally, 4′ thioIDU was shown to be protective against both cowpox and vaccinia in mice (Kern et al., 2009). Selectivity indices (CC50/EC50) ranged from more than 200 to 2,000 for 4′ thioIDU; in contrast, the values for CDV were more than 9 to more than 32 (Kern et al., 2009). However, 4′ thioIDU, like CDV and its derivatives, is toxic for dividing cells.

Agents That Enhance or Modulate the Host Immune Response

Enhancing or modulating the host immune response is an alternative or adjunctive therapeutic approach to controlling smallpox through antiviral drugs that disrupt the replication cycle. Providing passive immunity through the transfer of protective antibodies from an immune to a susceptible individual can lend temporary, but potentially life-saving, protection.

As an example, this approach was used therapeutically in the 1940s in Morocco. Antiserum was obtained from smallpox survivors soon after the last scabs fell off, and was then administered to newly arriving patients at the clinic in doses of 10–20 ml per day (Couzi and Kircher, 1941). Among the 200 persons given this treatment, including 75 patients with advanced hemorrhagic disease, all survived. However, this was a report of clinical experience, not a controlled study, and use of passive antibodies as therapy for clinically evident, established infection has not been demonstrated to be effective against systemic viral illnesses.

Today, VIG collected from individuals with high antibody titers from repeated immunization is given to confer passive immunity in individuals with complications resulting from smallpox vaccination, and is the only currently available intervention other than unlicensed antiviral drugs (Kempe et al., 1961; Wittek, 2006). Two intravenous formulations of VIG (Cangene and Dynport) have been licensed by the FDA for the management of patients with progressive vaccinia, eczema vaccinatum, severe generalized vaccinia, and extensive body surface involvement or periocular implantation of vaccinia following inadvertent inoculation (Wittek, 2006). When given to exposed individuals, VIG is expected to provide protection against infection for approximately 2–3 weeks, presumably through its neutralizing activity against vaccinia.

The conserved orthopox protein vaccinia B5/variola B6 is a major neutralizing target for VIG, although major neutralizing sites on B5 are exposed differently on the variola ortholog (Aldaz-Carroll et al., 2007). B5 is needed to wrap the MV to form extracellular virus, and interactions with actin are necessary for virion egress from the infected cell (see Aldaz-Carroll et al., 2005, 2007).

More recently, humanized chimpanzee monoclonal antibodies specific for the B5 and A33 envelope glycoproteins of vaccinia virus and the variola virus homologs have been reported to inhibit the spread of vaccinia and variola viruses in vitro and have conferred protection in a mouse model of poxvirus infection (Chen and Ron, 2006; Chen et al., 2007). These antibodies may be useful for treating vaccine-related complications or for prophylaxis or therapy of smallpox.

VECTOR reports that since 2002 it has been working to develop human recombinant antibodies as therapeutics for treatment of smallpox infection (Tikunova et al., 2005; Yun et al., 2006; Dubrovskaia et al., 2007). To that end, a panel of 66 unique human mini-antibodies against orthopoxviruses, including variola virus, was selected from VECTOR's combinatory phage library and from that obtained from The Medical Research Council (UK). Half of the antibodies selected were tested for their ability to neutralize variola virus. Based on the most promising antibodies, VECTOR states that four fully human antibodies against variola virus were constructed, their affinity constants were measured, and they were tested for their ability to neutralize vaccinia virus (personal communication, Ilya Drozdov, WHOCC for Orthopoxvirus Diagnosis and Repository for Variola Virus Strains and DNA, March 27, 2009).

Other antiviral drugs, such as ribavirin, that are not poxvirus specific but counteract host responses represent another therapeutic approach to smallpox infection (Baker et al., 2003). The lower specificity and potential toxicity of such drugs make them less ideal, but some have been powerful modulators of disease severity with life-saving effects. Two recent reviews have suggested that agents such as tumor necrosis factor (TNF) inhibitors that are used to treat septic shock may be effective (Harrison et al., 2004; Jahrling et al., 2005). However, recent transcriptome profiling studies (Rubins et al., 2004) have shown that virulent poxvirus infection in primates appears to suppress TNF expression, raising concerns that further TNF suppression may enhance virulence and produce more severe disease. Moreover, several laboratory studies have suggested that the TNF-inhibiting genes of the poxviruses are a crucial part of pathogenesis (Sedger et al., 2006; Bartee et al., 2009). TNF production therefore is likely to be a protective mechanism counteracted by orthopoxvirus proteins. This finding also suggests that disease would be exacerbated by anti-TNF treatment.

As noted, postexposure smallpox vaccination is beneficial if given shortly after the contact. In addition to accelerating the development of specific antiviral responses, vaccinia inoculation may elicit immediate innate responses that control the initial progression of infection and modulate disease severity. In a monkeypox model, however, postexposure vaccination was not as effective as cidofovir or its derivatives (Stittelaar et al., 2006).

Summary

At present, two drugs that are FDA-approved for other purposes—cidofovir, a DNA synthesis inhibitor, and Gleevec, a tyrosine kinase inhibitor—hold potential for use as therapeutics against smallpox. Cidofovir can be used on an investigational basis for treating severe orthopoxvirus infections, including smallpox. FDA-approved preparations of VIG are also available. New drugs that are under evaluation and show promise include orally bioavailable esters of cidofovir (CMX-001) and ST-246, an inhibitor of virus egress. ST-246 has been given to human volunteers and has been administered on a compassionate use therapeutic basis to a 2-year-old child with eczema vaccinatum following vaccinia exposure. New types of VIG are also being developed that target specific proteins such as variola B6R.

REGULATORY REQUIREMENTS

Recommendations and requirements for U.S. licensure of drugs intended for the prevention and treatment of variola infection are outlined extensively in a 2007 *Guidance for Industry* document prepared by the FDA (see also Chapter 1) (FDA, 2007). The guidance pertains primarily to small-molecule therapeutics, although its main principles can also be applied to biological products such as immunoglobulin preparations, monoclonal antibodies, and therapeutic proteins. Of particular relevance, demonstration of efficacy against live variola virus appears to be an essential step on the pathway to licensure (see Table 6-3). More specifically, use of the Animal

TABLE 6-3 Scientific Pathway for Drug Development

Steps	Assays	Criteria
1. Rational Design (optional)	Computerized displays of viral proteins and best fit of drugs	Drug fits target
2. Cell culture tests of effects of drugs on infected cells	Drug effects on cytotoxicity, virus production, cytopathy, comet formation, generation of resistant mutants	Efficacy/toxicity (EC50/CC50) >10
3. Small-animal model	Use of mice infected with ectromelia, cowpox, or vaccinia for initial studies of drug safety and efficacy in vivo	Doses and routes for treatment found where virus titers decrease by >3 logs, disease signs are eliminated in most animals, and mortality decreases >50%
4. Large-animal model	Cynomolgus macaques given monkeypox intratracheally or variola intravenously should be tested for shedding, virus titers, disease signs or A nonrodent model, such as one using rabbits or monkeys, should be tested for shedding, virus titers, transmission, disease signs	Same as above
5. Human beings	Safety trials in humans should monitor blood chemistries and other biomarkers for toxicity; infected people should also be tested for virus	• Phase I/II clinical trials • Treatment or emergency use Investigational New Drug (IND) application for severe vaccinia or other orthopoxvirus infections • Emergency Use Authorization (see Chapter 1)

Rule (see Chapter 1) or any other currently available regulatory pathway to achieve licensure is essentially precluded by the exceptionally narrow host range of variola virus; the lack of any previously recognized effective drug for use in head-to-head comparison with any new compound; and known and possible differences between variola and other orthopoxviruses in disease characteristics, drug susceptibility, and host range (Jordan and Hruby, 2006; Bolken and Hruby, 2008). Further, FDA officials have highlighted the critical importance of conducting safety studies in normal human volunteers and potentially in patients with underlying medical conditions. The FDA also recommends studies using animal models that mimic human disease progression to provide supporting evidence of clinical efficacy (Roberts et al., 2008), but the development of a nonhuman primate challenge model for variola has been extraordinarily difficult in practice (see Chapter 4); moreover, some orally administered candidates (e.g., CMX-001) are not absorbed in these animals. While data derived from studies of other orthopoxviruses (e.g., monkeypox or vaccinia) cannot be considered definitive evidence of antivariola activity, the FDA guidance indicates that exploratory studies with these viruses can provide important adjunctive information.

In addition to variola-specific considerations, general considerations applicable to the licensure of any antiviral agent include analysis of in vitro activity in conjunction with other drug candidates, selection and evaluation of resistant viral strains, and consideration of drug–vaccine and drug–drug interactions.

NEED FOR LIVE VARIOLA VIRUS

Fewer than 10 percent of published studies related to the development of therapeutics for smallpox have actually involved the use of live variola virus. This fact demonstrates that much can be accomplished by other means. For both scientific and regulatory reasons, however, the advanced stages of drug development will require evaluations involving live variola virus.

In the 30 years since the eradication of smallpox, variola stocks have been used to complete the sequence of at least 49 VARV isolates (see Chapter 5) (Esposito et al., 2006), to gain some understanding of the genetic differences between virulent and nonvirulent poxviruses, to understand the neutralizing epitopes that could be targeted by VIG, to further understand the replication cycle of variola in order to identify potential targets for antiviral agents, and to design and evaluate potential variola model challenge systems for purposes of confirming the efficacy of candidate antiviral agents under the Animal Rule.

Although preliminary testing of antivirals can use related orthopoxviruses, live variola virus should be used in cell culture as the ultimate test. Host cell responses that define the course of variola infection in cell culture or in vivo, such as changes in gene expression or changes in signaling pathways, miRNA, or secreted cytokines, should be investigated to identify networks of responses that could serve as biomarkers of inhibition of virus infection by a candidate drug. Host responses to similar viruses, such as monkeypox or vaccinia, could be used to identify biomarkers associated with virulent or benign infection. The changes in these profiles found to be associated with successful drug treatment in these models could be used as candidate biomarkers indicating poxvirus control and further evaluated in nonhuman primates infected with variola. Similarly, the pharmacokinetic and pharmacodynamic properties of the drug in these models need to reflect those in humans.

If such biological parameters can be validated, it may eventually be possible to use these measures in developing an alternative to testing with live variola virus. For example, a VIG formulation containing monoclonal antibodies to variola B6 could be tested in mice infected with a vaccinia virus expressing the orthologous vaccinia B5. An ectromelia infection of mice could perhaps have a profile similar to a variola infection of mice, resulting in the same alterations in host response whether the challenge virus was ectromelia or variola. Thus, host responses to drug treatments after ectromelia infection could serve as surrogate biomarkers for efficacy in the absence of live variola infection. Nevertheless, in accordance with Table 4-1 in Chapter 4, biomarkers developed in nonhuman primate models would be more likely to reflect the disease progression in human beings and therefore make better surrogates for disease progression in antiviral testing studies. Any predictions about drug activity against variola would have to be made with great caution.

REFERENCES

Aldaz-Carroll, L., J. C. Whitbeck, M. Ponce de Leon, H. Lou, L. Hirao, S. N. Isaacs, B. Moss, R. J. Eisenberg, and G. H. Cohen. 2005. Epitope-mapping studies define two major neutralization sites on the vaccinia virus extracellular enveloped virus glycoprotein B5R. *Journal of Virology* 79(10):6260–6271.

Aldaz-Carroll, L., Y. Xiao, J. C. Whitbeck, M. Ponce de Leon, H. Lou, M. Kim, J. Yu, E. L. Reinherz, S. N. Isaacs, R. J. Eisenberg, and G. H. Cohen. 2007. Major neutralizing sites on vaccinia virus glycoprotein B5 are exposed differently on variola virus ortholog B6. *Journal of Virology* 81(15):8131–8139.

Arumugham, B., H. J. Kim, M. N. Prichard, E. R. Kern, and C. K. Chu. 2006. Synthesis and antiviral activity of 7-deazaneplanocin A against orthopoxviruses (vaccinia and cowpox virus). *Bioorganic & Medicinal Chemistry Letters* 16:285–287.

Bailey, T. R., S. R. Rippin, E. Opsitnick, C. J. Burns, D. C. Pevear, M. S. Collett, G. Rhodes, S. Tohan, J. W. Huggins, R. O. Baker, E. R. Kern, K. A. Keith, D. Dai, G. Yang, D. Hruby, and R. Jordan. 2007. N-(3, 3a, 4, 4a, 5, 5a, 6, 6a-Octahydro-1,3-dioxo-4,6-ethenocytoprop[f]isoindol-2-(1H)-yl)carboxamides: Identification of novel orthopoxvirus egress inhibitors. *Journal of Medicinal Chemistry* 50:1442–1444.

Baker, R.O., M. Bray, and J. W. Huggins. 2003. Potential antiviral therapeutics for smallpox, monkeypox and other orthopoxvirus infections. *Antiviral Research* 57(1-2):13–23.

Bartee, E., M. R. Mohamed, M. C. Lopez, H. V. Baker, and G. McFadden. 2009. The addition of tumor necrosis factor plus beta interferon induces a novel synergistic antiviral state against poxviruses in primary human fibroblasts. *Journal of Virology* 83(2):498–511.

Becker, M. N., M. Obraztsova, E. R. Kern, D. C. Quenelle, K. A. Keith, M. N. Prichard, M. Luo, and R. W. Moyer. 2008. Isolation and characterization of cidofovir resistant vaccinia viruses. *Virology Journal* 5:58.

Bolken, T. C., and D. E. Hruby. 2008. Discovery and development of antiviral drugs for biodefense: Experience of a small biotechnology company. *Antiviral Research* 77(1):1–5.

Bond, A., Z. Reichert, and J. T. Stivers. 2006. Novel and specific inhibitors of a poxvirus type I topoisomerase. *Molecular Pharmacology* 69(2):547–557.

Bradbury, J. 2002. Orally available cidofovir derivative active against smallpox. *Lancet* 359:1041.

Buller, R. M., G. Owens, J. Schriewer, L. Melman, J. R. Beadle, and K. Y. Hostetler. 2004. Efficacy of oral active ether lipid analogs of cidofovir in a lethal mousepox model. *Virology* 318:474–481.

CDC (Centers for Disease Control and Prevention). 2009. Progressive vaccinia in a military smallpox vaccine United States, 2009. *MMWR. Morbidity and Mortality Weekly Report* 58(19):532–536.

Chen, C. C., and Y. Ron. 2006. New strategies for immune-mediated anti-viral drug and vaccine development. *Current Pharmaceutical Design* 12(11):1391–1401.

Chen, Z., P. Earl, J. Americo, I. Damon, S. K. Smith, F. Yu, A. Sebrell, S. Emerson, G. Cohen, R. J. Eisenberg, I. Gorshkova, P. Schuck, W. Satterfield, B. Moss, and R. Purcell. 2007. Characterization of chimpanzee/human monoclonal antibodies to vaccinia virus A33 glycoprotein and its variola virus homolog in vitro and in a vaccinia virus mouse protection model. *Journal of Virology* 81(17):8989–8995.

Ciesla, S. L., J. Trahan, W. B. Wan, J. R. Beadle, K. A. Aldern, G. R. Painter, and K. Y. Hostetler. 2003. Esterification of cidofovir with alkoxyalkanols increases oral bioavailability and diminishes drug accumulation in kidney. *Antiviral Research* 59(3):163–171.

Couzi, G., and J. P. Kircher. 1941. Immunotherapy of smallpox. *Bulletin de l'Institut d'Hygiene de Maroc* 1:59–68.

Da Fonseca, F., and B. Moss. 2003. Poxvirus DNA topoisomerase knockout mutant exhibits decreased infectivity associated with reduced early transcription. *Proceedings of National Academy of Sciences of the United States of America* 100:11291–11296.

De Clercq, E., and A. Holy. 2005. Acyclic nucleoside phosphonates: A key class of antiviral drugs. *Nature Reviews. Drug Discovery* 4:928–940.

Druker, B. J., S. Tamura, E. Buchdunger, S. Ohno, G. M. Segal, S. Fanning, J. Zimmermann, and N. B. Lydon. 1996. Effects of a selective inhibitor of the Abl tyrosine kinase on the growth of Bcr-Abl positive cells. *Nature Medicine* 2:561–566.

Dubrovskaia, V. V., A. B. Ulitin, A. G. Laman, I. P. Gileva, N. I. Bormotov, A. A. Il'ichev, F. A. Brovko, S. N. Shchelkunov, E. F. Belanov, and N. V. Tikunova. 2007. Construction of combinatorial immune library of single chain human antibodies to orthopoxviruses and selection from this library antibodies to recombinant protein prA30L of variola virus. *Molekuliarnaia Biologiia* 41(1):173–185.

Duraffour, S. R., M. Krecmerová, J. van Den Oord, R. De Vos, A. Holý, J. M. Crance, D. Garin, E. De Clercq, and G. Andrei. 2007. Activities of several classes of acyclic nucleoside phosphonates against camelpox virus replication in different cell culture models. *Antimicrobial Agents and Chemotherapy* 51:4410–4419.

Esposito, J. J., S. A. Sammons, A. M. Frace, J. D. Osborne, M. Olsen-Rasmussen, M. Zhang, D. Govil, I. K. Damon, R. Kline, M. Laker, Y. Li, G. L. Smith, H. Meyer, J. W. Leduc, and R. M. Wohlhueter. 2006. Genome sequence diversity and clues to the evolution of variola (smallpox) virus. *Science* 313:807–812.

Fan, X., X. Zhang, L. Zhou, K. A. Keith, E. R. Kern, and P. F. Torrence. 2006. Assembling a smallpox biodefense by interrogating 5-substituted pyrimidine nucleoside chemical space. *Antiviral Research* 71:201–205.

FDA (U.S. Food and Drug Administration). 2007. *Guidance for industry. Smallpox (variola) infection): Developing drugs for treatment or prevention.* http://www.fda.gov/cder/guidance/7423dft.pdf (accessed January 26, 2009).

Fenner, F., D. A. Henderson, I. Arita, J. Jezek, and L. D. Ladnyi. 1998. *Smallpox and its eradication*. Geneva, Switzerland: WHO.

Fujimoto, D. F., C. Pinilla, and A. M. Segall. 2006. New peptide inhibitors of type 1B topoisomerases: Similarities and differences vis-à-vis inhibitors of tyrosine recombinases. *Journal of Molecular Biology* 363:891–907.

Harrison, S. C., B. Alberts, E. Ehrenfeld, L. Enquist, H. Fineberg, S. L. McKnight, B. Moss, M. O'Donnell, H. Ploegh, S. L. Schmid, K. P. Walter, and J. Theriot. 2004. Discovery of antivirals against smallpox. *Proceedings of the National Academy of Sciences of the United States of America* 101(31):11178–11192.

Henderson, D. A., and F. Fenner. 2001. Recent events and observations pertaining to smallpox virus destruction in 2002. *Clinical Infectious Diseases* 33(7):1057–1059.

Hostetler, K. Y. 2009 (in press). Alkoxyalkyl prodrugs of acylic nucleoside phopspahtes enhance oral antiviral activity and reduce toxicity: Current state of the art. *Antiviral Research*.

Hostetler, K. Y., J. R. Beadle, J. Trahan, K. A. Aldern, G. Owens, J. Schwiewer, L. Melman, and M. R. Buller. 2007. Oral 1-O-octodecayl-2-O-benzyl-sn-glycero-3-cidofovir targets the lung and is effective against a lethal respiratory challenge with ectromelia virus in mice. *Antiviral Research* 73:212–218.

Ivanov, A. V., A. R. Simonian, E. F. Belanov, and L. A. Aleksandrova. 2005. Synthesis and antiviral activity of new 5-substituted 2′-deoxyuridine derivatives. *Bioorganicheskaia Khimiia* 31(6):616–622.

Ivanov, M. A., A. V. Ivanov, I. A. Krasnitskaia, O. A. Smirnova, I. L. Karpenko, E. F. Belanov, V. S. Prasolov, V. L. Tunitskaia, and L. A. Aleksandrova. 2008. New furano- and pyrrolo[2,3-d]pyrimidine nucleosides and their 5′-triphosphates: Synthesis and biological properties. *Bioorganicheskaia Khimiia* 34(5):661–670.

Jahrling, P. B., E. A. Fritz, and L. E. Hensley. 2005. Countermeasures to the bioterrorist threat of smallpox. *Current Molecular Medicine* 5(8):817–826.

Jordan, R., and D. Hruby. 2006. Smallpox antiviral drug development: Satisfying the animal efficacy rule. *Expert Review of Antiinfective Therapy* 4(2):277–289.

Jordan, R., D. Tien, T. C. Bolken, K. F. Jones, S. R. Tyavanagimatt, J. Strasser, A. Frimm, M. L. Corrado, P. G. Strome, and D. E. Hruby. 2008. Single-dose safety and pharmacokinetics of ST-246, a novel orthopoxvirus egress inhibitor. *Antimicrobial Agents and Chemotherapy* 52(5):1721–1727.

Kempe, C. H., C. Bowles, G. Meiklejohn, T. O. Berge, L. St Vincent, B. V. Babu, S. Govindarajan, N. R. Ratnakannan, A. W. Downie, and V. R. Murthy. 1961. The use of vaccinia hyperimmune gamma-globulin in the prophylaxis of smallpox. *Bulletin of the World Health Organization* 25:41–48.

Kern, E. R. 2003. In vitro activity of potential anti-poxvirus agents. *Antiviral Research* 57:35–40.

Kern, E. R., C. Hartline, E. Harden, K. Keith, N. Rodriguez, J. R. Beadle, and K. Y. Hostetler. 2002. Enhanced inhibition of orthopoxvirus replication in vitro by alkoalkyl esters of cidofovir and cyclic cidofovir. *Antimicrobial Agents and Chemotherapy* 46:991–995.

Kern, E. R., D. J. Collins, W. B. Wan, J. R. Beadle, K. Y. Hostetler, and D. C. Quenelle. 2004. Oral treatment of murine cytomegalovirus infections with ether lipid esters of cidofovir. *Antimicrobial Agents and Chemotherapy* 48:3516–3522.

Kern, E. R., M. N. Prichard, D. C. Quenelle, K. A. Keith, K. N. Tiwari, J. A. Maddry, and J. A. Secrist. 2009. Activities of certain 5′-substituted 4′-thiopyrimidine nucleosides against orthopoxvirus infections. *Antimicrobial Agents and Chemotherapy* 53(2):572–579.

Krecmerová, M., A. Holý, A. Pískala, M. Masojídková, G. Andreii, L. Naesens, J. Neyts, J. Balzarini, E. De Clercq, and R. Snoeck. 2007a. Antiviral activity of triazine analogues of 1-(S)-[3-hydroxy-2-(phosphonomethoxy)propyl]cytosine (cidofovir) and related compounds. *Journal of Medicinal Chemistry* 50:1069–1077.

Krecmerová, M., A. Holý, A. Pískala, M. Masojídková, G. Andreii, L. Naesens, J. Neyts, J. Balzarini, E. De Clercq, and R. Snoeck. 2007b. Ester prodrugs of cyclic 1-(S)-[3-hydroxy-2-(phosphonomethoxy)propyl]-5-azacytosine: Synthesis and antiviral activity. *Journal of Medicinal Chemistry* 50:5765–5772.

Lebeau, I., G. Andrei, F. Dal Pozzo, J. R. Beadle, K. Y. Hostetler, E. De Clercq, J. van den Oord, and R. Snoeck. 2006. Activities of alkoxyalkyl esters of cidofovir (CDV), cyclic CDV, and (S)-9-(3-hydroxy-2-phosphonylmethoxypropyl)adenine against orthopoxviruses in cell monolayers and in organotypic cultures. *Antimicrobial Agents and Chemotherapy* 50(7):2525–2529.

LeDuc, J. W., I. Damon, J. M. Meegan, D. A. Relman, J. Huggins, and P. B. Jahrling. 2002. Smallpox research activities: U.S. interagency collaboration, 2001. *Emerging Infectious Diseases* 8(7):743–745.

Magee, W. C., K. Y. Hostetler, and D. H. Evans. 2005. Mechanism of inhibition of vaccinia virus DNA polymerase by cidofovir diphosphate. *Antimicrobial Agents and Chemotherapy* 49:3153–3162.

Magee, W. C., K. A. Aldern, K. Y. Hostetler, and D. H. Evans. 2008. Cidofovir and (S)-9-[3-hydroxy-(2-phosphonomemethoxy)propyl]adenine are highly effective inhibitors of vaccinia virus DNA polymerase when incorporated into the template strand. *Antimicrobial Agents and Chemotherapy* 52:586–597.

McFadden, G. 2005. Gleevec casts a pox on poxviruses. *Nature Medicine* 11(7):711–712.

Morris, K. 2002. Oral drug and old vaccine renew smallpox bioterror debate. *Lancet Infectious Disease* 2:262.

Mortimer, P. P. 2003. Can postexposure vaccination against smallpox succeed? *Clinical Infectious Diseases* 36:222–229.

Naesens, L., G. Andrei, I. Votruba, M. Krecmerová, A. Holý, J. Neyts, E. De Clercq, and R. Snoeck. 2008. Intracellular metabolism of the new antiviral compound 1-(S)-[3-hydroxy-2-(phosphonomethoxy)propyl]-5-azacytosine. *Biochemical Pharmacology* 76(8):997–1005.

Nalca, A., J. M. Hatkin, N. L. Garza, D. K. Nichols, S. W. Norris, D. E. Hruby, and R. Jordan. 2008. Evaluation of orally delivered ST-246 as postexposure prophylactic and antiviral therapeutic in an aerosolized rabbitpox rabbit model. *Antiviral Research* 79:121–127.

Painter, G. R., and K. Y. Hostetler. 2004. Design and development of oral drugs for the prophylaxis and treatment of smallpox infection. *Trends in Biotechnology* 22(8):423–427.

Parker, S., E. Touchette, C. Oberle, M. Almond, A. Robertson, L. C. Trost, B. Lampert, G. Painter, and R. M. Buller. 2008. Efficacy of therapeutic intervention with an oral ether-lipid analogue of cidofovir (CMX001) in a lethal mousepox model. *Antiviral Research* 77(1):39–49.

Parker, S., A. M. Siddiqui, C. Oberle, E. Hembrador, R. Lanier, G. Painter, A. Robertson, and R. M. Buller. 2009. Mousepox in the C57BL/6 strain provides an improved model for evaluating antipoxvirus therapies. *Virology* 385(1):11–21.

Perry, K., Y. Hwang, F. D. Bushman, and G. D. Van Duyne. 2006. Structural basis for specificity in the poxvirus topoisomerase. *Molecular Cell* 23:343–354.

Prichard, M. N., K. A. Keith, M. P. Johnson, E. A. Harden, A. McBrayer, M. Luo, S. Qiu, D. Chattopadhyay, X. Fan, P. F. Torrence, and E. R. Kern. 2007. Selective phosphorylation of antiviral drugs by vaccinia virus thymidine kinase. *Antimicrobial Agents and Chemotherapy* 51:1795–1803.

Quenelle, D. C., D. J. Collins, and E. R. Kern. 2003. Efficacy of multiple- or single-dose cidofovir against vaccinia and cowpox virus infections in mice. *Antimicrobial Agents and Chemotherapy* 47:3275–3280.

Quenelle, D. C., R. M. Buller, S. Parker, K. A. Keith, D. E. Hruby, R. Jordan, and E. R. Kern. 2007a. Efficacy of delayed treatment with ST-246 given orally against systemic orthopoxvirus infections in mice. *Antimicrobial Agents and Chemotherapy* 51:689–695.

Quenelle, D. C., M. N. Prichard, K. A. Keith, D. E. Hruby, R. Jordan, G. R. Painter, A. Robertson, and E. R. Kern. 2007b. Synergistic efficacy of the combination of ST-246 with CMX001 against orthopoxviruses. *Antimicrobial Agents and Chemotherapy* 51:4118–4124.

Reeves, P. M., B. Bommarius, S. Lebeis, S. McNulty, J. Christensen, A. Swimm, A. Chahroudi, R. Chavan, M. B. Feinberg, D. Veach, W. Bornmann, M. Sherman, and D. Kalman. 2005. Disabling poxvirus pathogenesis by inhibition of Abl-family tyrosine kinases. *Nature Medicine* 11(7):731–739.

Reeves, P. M., S. K. Smith, V. A. Olson, S. McNulty, I. Damon, and D. Kalman. 2006. *Potential utility of host tyrosine kinase inhibitors as therapeutics for poxvirus infections*. Abstract presented at XVI International Poxvirus and Iridovirus Symposium, Palm Springs, California.

Roberts, R., B. Styrt, and S. McCune. 2008. FDA perspective on antivirals against biothreats: Communicate early and often. *Antiviral Research* 78(1):60–63.

Roy, A., S. W. Schneller, K. A. Keith, C. B. Hartline, and E.R. Kern. 2005. The 4′,4′-difluoro analog of 5′-noraristeromycin: a new structural prototype for possible antiviral drug development toward orthopoxvirus and cytomegalovirus. *Bioorganic & Medicinal Chemistry* 13 (14):4443–4449.

Rubins, K. H., L. E. Hensley, P. B. Jahrling, A. R. Whitney, T. W. Geisbert, J. W. Huggins, A. Owen, J. W. Leduc, P. O. Brown, and D. A. Relman. 2004. The host response to smallpox: analysis of the gene expression program in peripheral blood cells in a nonhuman primate model. *Proceedings of the National Academy of Sciences of the United States of America* 101(42):15190–15195.

Ruiz, J. C., J. R. Beadle, K. A. Aldern, K. A. Keith, C. B. Hartline, E. R. Kern, and K. Y. Hostetler. 2007. Synthesis and antiviral evaluation of alkoxyalkyl-phosphate conjugates of cidofovir and adefovir. *Antiviral Research* 75:87–90.

Sbrana, E., R. Jordan, D. E. Hruby, R. I. Mateo, S. Y. Xiao, M. Siirin, P. C. Newman, A. P. Da Rosa, and R. B. Tesh. 2007. Efficacy of the antipoxvirus compound ST-246 for treatment of severe orthopoxvirus infection. *The American Journal of Tropical Medicine and Hygiene* 76:768–773.

Sedger, L. M., S. R. Osvath, X. M. Xu, G. Li, F. K. Chan, J. W. Barrett, and G. McFadden. 2006. Poxvirus tumor necrosis factor receptor (TNFR)-like T2 proteins contain a conserved preligand assembly domain that inhibits cellular TNFR1-induced cell death. *Journal of Virology* 80(18):9300–9309.

Sliva, K., and B. Schnierle. 2007. From actually toxic to highly specific—novel drugs against poxviruses. *Journal of Virology* 4:8.

Stittelaar, K. J., J. Neyts, L. Naesens, G. van Amerongen, R. F. van Lavieren, A. Holý, E. De Clercq, H. G. Niesters, E. Fries, C. Maas, P. G. Mulder, B. A. van der Zeijst, and A. D. Osterhaus. 2006. Antiviral treatment is more effective than smallpox vaccination upon lethal monkeypox virus infection. *Nature* 439(7077):745–748.

Tesh, R. B., D. M. Watts, E. Sbrana, M. Siirin, V. L. Popov, and S. Y. Xiao. 2004. Experimental infection of ground squirrels (spermophilius tridecemlineatus) with monkeypox virus. *Emerging Infectious Diseases* 10(9):1563–1567.

Tikunova, N. V., E. I. Bovshik, T. E. Iun, E. V. Zhirakovskaia, V. V. Morozova, T. A. Batanova, A. A. Gus'kov, E. B. Sokunova, A. A. Il'ichev, and L. S. Sandakhchiev. 2005. Human recombinant antibodies to variola virus. *Voprosy Birusologii* 50(6):20–25.

Tse-Dinh, Y. C. 2008. An update on the development of drugs against smallpox. *Current Opinion in Investigational Drugs* 9:865–870.

Vora, S., I. Damon, V. Fulginiti, S. G. Weber, M. Kahana, S. L. Stein, S. I. Gerber, S. Garcia-Houchins, E. Lederman, D. Hruby, L. Collins, D. Scott, K. Thompson, J. V. Barson, R. Regnery, C. Hughes, R. S. Daum, Y. Li, H. Zhao, S. Smith, Z. Braden, K. Karem, V. Olson, W. Davidson, G. Trindade, T. Bolken, R. Jordan, D. Tien, and J. Marcinak. 2008. Severe eczema vaccinatum in a household contact of a smallpox vaccinee. *Clinical Infectious Diseases* 46:1555–1561.

Whitley, R. 2008. Development of Small Molecule Therapies of Orthopoxvirus Replication. Presentation to Committee, December 2008. Washington, DC.

Wittek, R. 2006. Vaccinia immune globulin: Current policies, preparedness, and product safety and efficacy. *International Journal of Infectious Diseases* 10(3):193–201.

Yang, G., D. C. Pevear, M. H. Davies, M. S. Collett, T. Bailey, S. Rippen, L. Barone, C. Burns, G. Rhodes, S. Tohan, J. W. Huggins, R. O. Baker, R. L. M. Buller, E. Touchette, K. Waller, J. Schriewer, J. Neyts, E. DeClercq, K. Jones, D. Hruby, and R. Jordan. 2005. An orally bioavailable antipoxvirus compound (ST-246) inhibits extracellular virus formation and protects mice from lethal orthopoxvirus challenge. *Journal of Virology* 79:13139–13149.

Yang, M., and S. W. Schneller. 2005. 5'-Homoaristeromycin. Synthesis and antiviral activity against orthopox viruses. *Bioorganic & Medicinal Chemistry Letters* 15 (1):149–151.

Yun, T. E., N. V. Tikunova, L. N. Shingarova, T. K. Aliev, E. F. Boldyreva, V. V. Morozova, A. N. Shvalov, O. V. Nekrasova, I. V. Polykhalova, A. A. Panina, A. A. Il'ichev, M. P. Kirpichnikov, and L. S. Sandakhchiev. 2006. The full-length recombinant human antibody to vaccinia virus. *Doklady. Biochemistry and Biophysics* 407:98–101.

Zakirova, N. F., A. V. Shipitsyn, E. F. Belanov, and M. V. Jasko. 2004. A new approach to the synthesis of optically active alkylated adenine derivatives. *Bioorganic & Medicinal Chemistry Letters* 14(12):3357–3360.

7

Development of Vaccines

Widespread vaccination was key to the success of the smallpox eradication program. Because smallpox no longer circulates as a natural pathogen, routine vaccination of the general population has been discontinued. In the event of an outbreak, however, vaccination still would be the most effective means of preventing an epidemic, and current U.S. guidelines call for vaccination of selected persons for preparedness purposes. WHO has a vaccine stockpile and recommends that national governments maintain stores of smallpox vaccine.

The 1999 IOM committee identified the need to maintain stores of smallpox vaccine:

> **Adequate stocks of smallpox vaccine would have to be maintained if research were to be conducted on variola virus or if maintenance of a smallpox vaccination program were required. Live variola virus would be necessary if certain approaches to the development of novel types of smallpox vaccine were to be pursued.**

The 1999 committee noted that, at the time, stocks of live vaccinia virus were deteriorating and would likely need to be replenished. However, the committee determined that live variola virus would not be required for the development of live vaccinia vaccines based on traditional vaccine strains but produced with modern tissue culture techniques, as safety and efficacy could be measured against the parental vaccine produced in animals. The committee concluded that research involving live variola virus would be required for the development of nonreplicating virus, live-attenuated virus, and subunit vaccines. The committee noted that such testing would also require as yet undeveloped animal models.

This chapter examines in turn the history and current status of smallpox vaccine development, the scientific pathway to development, salient regulatory requirements, and the need for live variola virus in this work.

HISTORY OF SMALLPOX VACCINE DEVELOPMENT

Early attempts at preventing smallpox arose from the observations that people who survived the disease had lifetime protection against new exposures and that those who were exposed to variola virus via the cutaneous route had milder disease. To prevent life-threatening smallpox, attempts were made to expose individuals to the virus from scabs, a process known as variolation. In 1796, Edward Jenner reported that milkmaids who had cowpox were immune to smallpox, as were children inoculated with material taken from cowpox lesions (see Figure 7-1). The practice of "vaccination" subsequently became widespread in Europe and North America during the nineteenth century.

Early Vaccine Development

While standards for smallpox vaccine composition and delivery were initially lacking, by the 1960s WHO had established guidelines mandating a specific concentration (1×10^8 plaque-forming units per milliliter) and a specific method (multiple puncture with a bifurcated needle) for smallpox vaccination. Delivery via the bifurcated needle involves dipping the needle in a suspension of vaccinia virus and repeatedly pricking the skin. The strain used in the vaccine varied by country, and certain strains were less reactogenic than others, but WHO's Intensified Smallpox Eradication Programme most often used the Lister strain (Parrino and Graham, 2006).

Posteradication Era

When smallpox threatened as many as one in four individuals, the risk of adverse effects of vaccination was less than the risk of contracting the disease itself. In the posteradication era, however, concerns about vaccine safety take precedence. The growing number of immunocompromised individuals globally has also changed the risk/benefit calculations for any future widespread use of traditional vaccinia vaccines (Artenstein and Grabenstein, 2008).

Until recently, Dryvax®, made from the NYCBH vaccinia strain, was the only smallpox vaccine licensed in the United States. Its efficacy against variola was established before eradication. However, use of live vaccinia

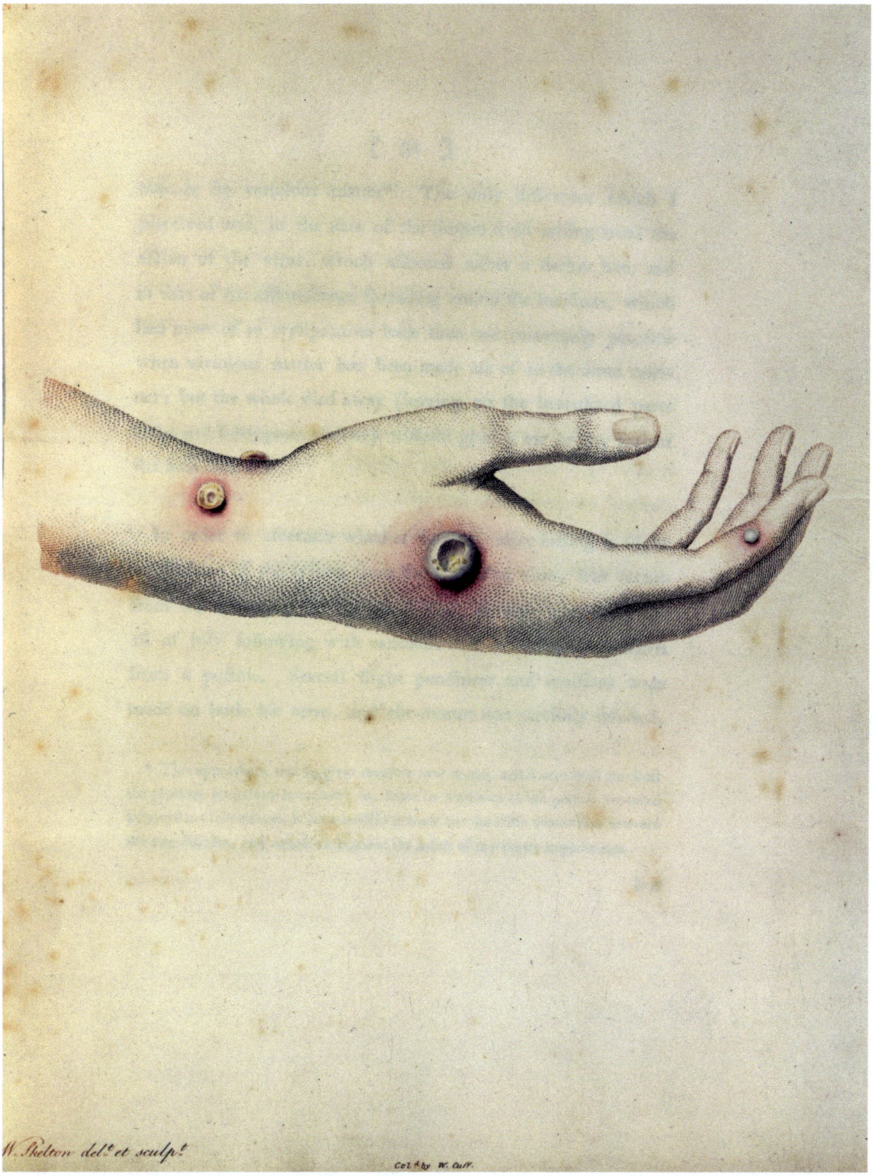

FIGURE 7-1 Cowpox pustule on the arm of Sarah Nelmes, from *An Inquiry into the Causes and Effects of the Variolae Vaccinae* by Edward Jenner (1749–1823), engraved by Pearce, ca 1800 (colored engraving) by William Skelton (1763–1848), Bibliotheque de la Faculté de Médecine, Paris, France/Archives Charmet/The Bridgeman Art Library.

virus can result in adverse effects, such as generalized and progressive vaccinia, eczema vaccinatum (EV), and postvaccinial encephalitis (see the discussion of first-generation vaccines below). Based on historical data obtained from two large-scale population-based surveys conducted through 1968, the risk of severe adverse effects is estimated to be around 1 in 1,000 vaccines; the risk of life-threatening or fatal effects is much lower (IOM, 2005).

U.S. Experience

In the wake of the 2001 terrorist attacks in the United States and the deaths due to anthrax that occurred soon after, national security and public health officials began to debate the adoption of a smallpox vaccination program to prepare the country for an intentional release of variola virus (Fauci, 2002; Seiler et al., 2003). In December 2002, the U.S. Government announced the Smallpox Vaccination Program, which involved the immediate and mandatory vaccination with Dryvax of up to 500,000 military personnel and the voluntary vaccination of up to 500,000 front-line health care workers and other critical personnel; this was to be followed by the vaccination of up to 10 million first responders, with plans to make the vaccine available to members of the general public (Wilson, 2005). By October 2003, however, the broader program had effectively ended (Wilson, 2005). Four factors led to this outcome: the lack of trained personnel and funds in state health departments charged with implementation (IOM, 2005), the lack of an injury compensation program (Seiler et al., 2003; Wilson, 2005), the emergence of reports of unanticipated cardiac events (Wilson, 2005), and waning concern about an intentional release of variola virus (Seiler et al., 2003; Chapman et al., 2008). At the request of CDC, the IOM produced a series of letter reports and a final summary on issues related to implementation of the vaccination program (IOM, 2003a–e, 2004). This experience provided important contemporary information on challenges in vaccination implementation absent endemic disease, as well as the safety and immunogenicity of smallpox vaccine.

No deaths attributable to the vaccine occurred among the military personnel (730,580) and civilians (39,566) given Dryvax during the Smallpox Vaccination Program (Poland et al., 2005; Chapman et al., 2008). Numbers of reports of anticipated adverse events were similar to or lower than those in the past (Strikas et al., 2008). One military and one civilian vaccinee developed encephalitis, while 50 military personnel developed generalized vaccinia (Lewis et al., 2006). There were 112 cases of inadvertent infection, 78 in vaccinees (autoinoculation) and 52 in close contacts of vaccinees (Poland et al., 2005). Myocardial ischemia was the most notable of the unanticipated adverse events, occurring in 24 military and 10 civilian vaccinees (Swerdlow et al., 2008). Five vaccinees, all of whom had preexisting heart disease, experienced fatal myocardial infarctions (Poland et al., 2005; Neff et al., 2008; Swerdlow et al., 2008); this rate did not exceed that expected among unvaccinated people with similar medical histories (Neff et al., 2008). Myopericarditis was diagnosed in 107 cases (86 military, 21 civilian) (Casey et al., 2005; Poland et al., 2005; Morgan et al., 2008), a significant increase associated with vaccinia vaccination (Neff et al., 2008; Strikas et al., 2008). This experience suggests that Dryvax and related products should not be given to individuals with known heart disease in the absence of a smallpox outbreak. CDC's adverse event reporting system remains in place as a means to further assess rare or unexpected complications of vaccinia vaccination (Thomas et al., 2008).

CURRENT STATUS OF SMALLPOX VACCINE DEVELOPMENT

The development of smallpox vaccines has progressed in three major phases, and vaccines are classified as first-, second-, or third-generation (see Table 7-1).

First Generation

Traditional or first-generation smallpox vaccines were used during the eradication program. These vaccines, made using vaccinia virus, are Jennerian vaccines, defined as live viral vaccines that are attenuated by virtue of their host range specificity. Vaccinia causes a small infection of the skin at the vaccination site, called a "take," which is the only known correlate of vaccine efficacy. These traditional vaccines were manufactured by growing the vaccinia virus in live animals, such as cattle and sheep (Collier, 1955, 1980). Dryvax is the only first-generation smallpox vaccine licensed in the United States (Artenstein and Grabenstein, 2008), while the Lister/Elstree vaccine is available in Europe. These are lyophilized preparations of live vaccinia virus prepared from calf lymph. The vaccines are made by inoculating animals with seed virus derived from the NYCBH (for Dryvax) or Elstree strain of vaccinia.

Although not subjected to any modern systematic scientific evaluation using live variola virus, the traditional vaccines set a benchmark against which all other smallpox vaccines must be measured because their efficacy has been established in the human population during natural outbreaks of smallpox. While the efficacy profile of first-generation vaccines is not completely known, the experience during eradication indicates a high level of effectiveness and infrequent serious adverse effects (IOM, 2005). In those who are immunocompromised and those who suffer from certain exfoliative skin conditions, however, the vaccinia virus can cause progres-

TABLE 7-1 Vaccines and Strains Used

Platform	Product	Parent Strain	Rationale for Use
		First Generation	
Lymph-derived vaccinia	Dryvax® (Wyeth)	NYCBH	Historical experience in the United States through the era of routine use
	Sanofi Pasteur smallpox vaccine (SPSV)	NYCBH	Produced in 1956–1957 and used in the U.S. program of that era; in frozen storage since
	Elstree-RIVM (master seed stock held at the National Institute of Public Health in The Netherlands [RIVM])	Lister	Historical experience in the Intensified Smallpox Eradication Programme
		Second Generation	
Replication-competent tissue-cultured vaccinia virus	ACAM2000™ (Acambis): cloned virus grown in Vero cells	NYCBH	Defined manufacturing process; reduced theoretical risk of adventitious agents compared with lymph-derived vaccine; less neurovirulent in animal models
	Elstree-BN (Bavarian-Nordic)	Lister	Defined manufacturing process; reduced theoretical risk of adventitious agents compared with lymph-derived vaccine
		Third Generation	
Replication-competent, highly attenuated vaccinia virus	LC16m8 vaccine: derived from 53 serial passages in rabbit kidney cells; temperature-sensitive, small-plaque phenotype due to mutation in the B5R gene	Lister	Experience in more than 100,000 Japanese children between 1973 and 1975; better safety profile than traditional live vaccinia, less neurovirulent in animals but unproven clinical efficacy

TABLE 7-1 Continued

Platform	Product	Parent Strain	Rationale for Use
Replication-deficient, highly attenuated vaccinia virus	MVA, derived from more than 570 serial passages in chicken embryo fibroblasts: IMVAMUNE (Bavarian-Nordic); TBC-MVA (Therion)	Ankara	Theoretically improved safety profile, especially for those in whom live vaccinia is contraindicated; used in 120,000 primary vaccinees in Germany in 1970s but unproven clinical efficacy
	NYVAC (Sanofi-Pasteur): attenuated by the deletion of 18 open reading frames from a plaque-cloned vaccinia isolate	Copenhagen	Theoretically improved safety profile, especially for those in whom live vaccinia is contraindicated
Subunit vaccines	Recombinant proteins; plasmid DNA	Vaccinia viruses, different sources	Theoretically improved safety profile

SOURCE: Adapted from Artenstein and Grabenstein, 2008.

sive or necrotizing vaccinia and EV, respectively. Progressive vaccinia is generally fatal, while EV is life-threatening. Moreover, because the vaccine site contains infectious virus, vaccinia can be transmitted to close contacts, putting these people unintentionally at risk (Lane et al., 1969; Fenner et al., 1988).

EV has long been one of the most serious adverse effects of vaccinia vaccination. It occurs in people with atopic dermatitis (AD), a condition associated with skin barrier dysfunction and defects in antiviral immunity (Wollenberg and Enger, 2004). The estimated incidence of EV in primary vaccinees is 40 in 1,000,000 (CDC). EV is characterized by extensive vaccina growth at the inoculation site or at the area affected by eczema. A recent case of life-threatening EV occurred in a child with AD who became infected by household contact with his father, who had been vaccinated against smallpox (Vora et al., 2008) (see also Chapter 6). With the increased incidence of AD, the potential risk of EV and its dire consequences in primary vaccinees and their contacts with AD cannot be underestimated (Horii et al., 2007).

The pathogenesis of EV is not completely understood, but important scientific advances have occurred since the 1999 IOM report was issued.

Keratinocytes are the predominant cell type in the epidermis. Liu and colleagues (2005) reported that vaccinia virus had limited replicative capacity in human keratinocytes and that infection induced keratinocytes to produce Th2 cytokines. Howell and colleagues (2004) showed that cathelicidin, an antimicrobial peptide produced by injured or infected skin, reduces vaccinia infectivity. Cathelicidin-deficient mice developed larger and more numerous skin lesions when infected by scarification with vaccinia virus. It was found that cathelicidin production rises in response to vaccinia infection of skin biopsies, and this response is attenuated in vaccinia-infected AD skin (Howell et al., 2006). Deng and colleagues (2008) reported that infection with a mutant vaccinia virus, ΔE3L (in which dsRNA-binding protein E3L is deleted) could be sensed by keratinocytes through an MAVS- and IRF3-dependent cytosolic RNA-sensing pathway to trigger the production of interferon and proinflammatory cytokines and chemokines. Further studies are needed to determine whether the skin of individuals with AD is deficient in mounting interferon responses to vaccinia infection.

In the modern era, from the late nineteenth century through global eradication, the development of first-generation vaccines was driven by concerns about both safety, with the aim of minimizing the reactogenicity of vaccines, and efficacy, manifested by the take rate, which served as a readily quantifiable correlate of efficacy. Low take rates in vaccine lots were generally ascribed to problems with production. During the eradication campaign, WHO addressed these concerns by acting to improve production processes in member nations and setting minimum standards for vaccine concentration as assessed by pock formation on chorioallantoic membrane (CAM), the heat stability of vaccine lots, and provision of standardized seed lots from the WHO collaborating centers (Fenner et al., 1988, Chapter 11). Consequently, live variola virus itself was not central to the development of first-generation vaccines beyond the original observations of Jenner himself and his immediate followers.

Second Generation

Because of the relatively high incidence of mild complications associated with tradiational vaccinia vaccines and the risk of severe complications in people with certain preexisting medical conditions, alternative vaccines are desirable. In addition, the use of live animals for production is inconsistent with modern pharmaceutical manufacturing practices and raises a theoretical concern about the spread of transmissible spongiform encephalopathies (TSEs), such as bovine spongiform encephalopathy (BSE)/mad cow disease.

The second-generation vaccines use live replicating vaccinia virus, but are produced using modern tissue culture techniques rather than growth in live animals. In a notable achievement since the first IOM report was issued, a second-generation vaccine, ACAM2000™, was recently licensed for use against smallpox by the FDA and has been added to the U.S. Strategic National Stockpile. The second-generation and first-generation vaccines are similar, but the former are more acceptable under the modern regulatory framework and avoid the potential hazards associated with TSEs. The same strains used in first-generation vaccines can also be used in second-generation vaccines. For example, ACAM2000 uses the NYCBH strain of vaccinia, prepared in Vero cells (Frey et al., 2009). This means that vaccines derived from tissue culture should bear a strong similarity to first-generation vaccines in terms of efficacy, but therefore also have the potential to cause the same spectrum of complications in both healthy recipients and those with medical contraindications, as well as in contacts of recipients accidentally infected.

In double-blind randomized trials assessing probable efficacy and safety, no significant differences in response (take rates and rates of adverse effects) were seen between ACAM2000 and Dryvax (Artenstein and Grabenstein, 2008; Frey et al., 2009). The clinical safety data on ACAM2000 suggest a continued risk of myopericarditis. The rate of myopericarditis in the Dryvax group is higher than that reported by the earlier U.S. Department of Defense (DOD) or CDC programs, but neither program had active surveillance in place for this particular adverse event (Greenberg and Kennedy, 2008). Another second-generation vaccine, CCSV, derived from cell culture, also showed a good safety profile in initial tests; however, there are no further plans to develop this vaccine (Bonilla-Guerrero and Poland, 2003; Artenstein and Grabenstein, 2008). ACAM2000 is licensed only for use in the Strategic National Stockpile. Live variola virus was not required for licensure of the second-generation ACAM2000 vaccine.

VECTOR reports production of a recombinant and highly attenuated strain of vaccinia virus, b7, 5S2-S, by the insertion of a hepatitis B (HB) DNA fragment into the thymidine kinase gene of vaccinia virus strain, L-IVP, coding for synthesis of the HBs and preS2-S proteins (Russian Federation Patent #1575576). Currently, based on this strain, a second-generation bivalent egg-based smallpox vaccine for oral administration is being developed (Russian Federation Patent #2076735). Increased safety of such a vaccine for the organism as compared with cutaneous smallpox vaccination arises from the switching off of the thymidine kinase gene of vaccinia virus that results from inserting the DNA fragment of HB virus. This vaccine has reportedly passed preclinical studies and Phase I clinical trials in a group of 100 subjects (Sergeev et al., 2004; Pliasunov et al., 2006; personal communication, Ilya Drozdov, WHOCC for Orthopoxvirus Diagnosis and Repository for Variola Virus Strains and DNA, March 27, 2009).

In addition, VECTOR is developing highly attenuated variants of live vaccines based on vaccinia virus using direct deletion of several genes, as well as DNA vaccines against smallpox. Such developments are in the preclinical phase (Maksyutov et al., 2006; personal communication, Ilya Drozdov, WHOCC for Orthopoxvirus Diagnosis and Repository for Variola Virus Strains and DNA, March 27, 2009).

Third Generation

The development of safer smallpox vaccines is necessary because of the adverse events associated with first-generation vaccines; as noted, second-generation vaccines resemble first-generation vaccines in that they contain infectious vaccina. In light of those adverse events, it has been estimated that at least 25 percent of the U.S. population should not receive traditional smallpox vaccines in the absence of an outbreak (Kemper et al., 2002). Smallpox vaccines that have an improved or potentially improved safety profile with respect to complications are often referred to as third- or next-generation smallpox vaccines, and can be subdivided into three distinct groups: nonreplicating virus, live-attenuated virus, and subunit vaccines.

Candidate **nonreplicating virus vaccines** include vaccinia virus derivates such as MVA (modified Vaccinia Ankara) and Nyvac (Hochstein-Mintzel et al., 1975; Tartaglia et al., 1992; Mahnel and Mayr, 1994; Paoletti et al., 1994). These are viruses that replicate in tissue culture but cannot replicate effectively in a human host or in immunocompromised animals. This vastly improves their safety profile, although there are probable increased production costs relative to the second-generation vaccines. While the leading candidate, MVA, was previously used widely in humans in the former West Germany in the 1970s (Mayr et al., 1978), variola virus was not endemic in West Germany at that time, and therefore no clinical data exist on MVA's effectiveness against smallpox.

MVA has a good safety profile and has been evaluated extensively as a third-generation smallpox vaccine. It was originally derived from vaccinia strain Ankara by more than 570 serial passages of the virus in primary chicken embryo fibroblasts (CEFs). MVA is highly attenuated and cannot replicate in humans and most mammalian cells. It has a 31 kilobase pair deletion from its parental genome and lacks several of the immunomodulatory gene products, such as soluble receptors for IFN-α, β, and γ; tumor necrosis factor; and CC chemokines. It also lacks proteins that affect host range and NF-κB signaling, such as K1L and A52R (Meyer et al., 1991; Antonie et al., 1998; Blanchard et al., 1998).

MVA infection of human monocyte-derived dendritic cells (DCs) increases the surface expression of costimulatory molecules and has a moderate induction of proinflammatory cytokines, whereas wild-type vaccinia strains do not (Drillien et al., 2004). Waibler and colleagues (2007) reported that MVA induces IFN-α in murine plasmacytoid DCs via a largely toll-like receptor (TLR)-independent mechanism. Samuelsson and colleagues (2008) demonstrated that MVA induces IFN-α in murine conventional DCs and plasmacytoid DCs via both TLR9-dependent and independent mechanisms. In a murine model, vaccination with MVA intranasally at the same time as or 2 days after a lethal dose of ectromelia virus (ECTV) protected the animals from death. MVA-mediated protection is partially dependent on the type I interferon receptor. These results provide some basis for the immunogenicity of MVA as a vaccine and suggest that it might be useful against a lethal poxvirus infection in a postexposure setting.

Mice with severe combined immunodeficiency can tolerate a 1,000 times higher dose of MVA than the standard vaccine strain (Dryvax) (Wyatt et al., 2004). Mice developed virus-specific $CD8^+$ T cells and neutralizing antibodies after MVA inoculation, and vaccinated mice were protected against lethal intranasal challenge of vaccinia WR strain. Mice deficient in B cells or $CD8^+$ T cells were also protected, whereas CD4 or MHC Class II knockout mice were poorly protected (Wyatt et al., 2004).

Extensive studies conduced in nonhuman primate models have demonstrated the efficacy of MVA against lethal monkeypox infection. MVA is safe in immune-deficient nonhuman primates (Stittelaar et al., 2001). Immunization with two doses of MVA alone or one dose of MVA followed by Dryvax generated neutralizing antibodies and antiviral-specific T cell responses equivalent to or higher than those induced by Dryvax alone and provided protection against an intravenous lethal challenge with monkeypox in a nonhuman primate model (Earl et al., 2004). Protection against respiratory challenges with monkeypox virus via the intratracheal route has also been shown with MVA (Stittelaar et al., 2005). MVA leads to a more rapid immune response than Dryvax in nonhuman primates. MVA administration 4 days prior to intravenous challenge with monkeypox provided protection, whereas Dryvax did not (Earl et al., 2008).

The results of the above studies suggest that MVA is safe and may be effective against smallpox (Mahnel, 1985; McCurdy et al., 2004; Coulibaly et al., 2005; Meseda et al., 2005; Slifka, 2005; Belyakov et al., 2006; Phelps et al., 2007; Damon et al., 2009). MVA vaccines are currently under development (Vollmar et al., 2006; Parrino et al., 2007).

Live-attenuated virus vaccines that retain limited ability to replicate in human hosts offer another route to a safer smallpox vaccine that may be appropriate for use in those for whom second-generation vaccines pose too high a risk. These vaccines, such as LC16m8, are more similar than nonreplicating virus vaccines to the second-generation vaccines by virtue of their ability to replicate in the vaccinee, but reduce the risk of complications. The LC16m8 strain is derived from the Lister/Elstree traditional vaccine strain and was used in Japan in the 1970s (Yamaguchi et al., 1975; Kidokoro et al., 2005), although, as with MVA in West Germany, smallpox was not endemic in Japan at that time.

Animal studies have shown that LC16m8 can protect monkeys from lethal monkeypox infection (Saijo et al., 2006). LC16m8 has also been shown to be nonlethal with no signs of disease in highly immunocompromised severe combined immunodeficient (SCID) mice (Kidokoro et al., 2005). In a trial involving monkeypox challenge in a nonhuman primate model, LC16m8 was tested for protective immunity in comparison with a live vaccinia vaccine derived from the Lister strain. Here, immunity conferred to both intranasal and subcutaneous challenge with monkeypox virus was equivalent in both groups, and greater than that of a nonimmunized group (Saijo et al., 2006). LC16m8 is currently licensed for use in Japan. Among 8,544 people who received LC16m8, the following adverse events occurred: 8 cases of urticaria, 1 mild case of EV, 9 cases of autoinoculation, 28 cases of rash localized around the vaccination site, and 3 benign febrile seizures (reviewed in Kenner et al., 2006). During 2002–2005, 1,529 members of the Japan Self-Defence Forces were vaccinated intraepidermally with Lc16m8, and 1,692 members were revaccinated. Fully 94 percent of the previously unvaccinated individuals presented a take, as did 86 percent of the revaccinated individuals. In addition, 200 of the subjects were tested for seroconversion; 96 percent of unvaccinated members and 60 percent of revaccinated individuals exhibited seroconversion or a booster response. No serious adverse events were reported; one case of allergic dermatitis and one case of erythema multiforme were observed.

Protein-based **subunit vaccines** do not contain genetic material and therefore cannot cause an infectious disease in the recipient. A potentially negative feature of these vaccines is that they contain only a limited number of the antigens of the target pathogen, and thus may induce a narrower immune response than a vaccine based on a whole virus. In addition, because these vaccines do not actively produce proteins in the vaccine recipient, the immune response induced is qualitatively different from that elicited by a live, nonreplicating or live-attenuated virus. Nonetheless, subunit vaccines based on up to three or four variola or vaccinia proteins have yielded promising results in the laboratory (Galmiche et al., 1999; Fogg et al., 2004), providing in animal models protection close to that of traditional vaccines in the short term. Subunit vaccine approaches that use a small quantity of DNA (around 1 percent) of the genome of variola or vaccinia virus allow active protein production in the recipient in a manner analogous to that of the replication-defective vaccines described above, and these approaches also have shown promise in the laboratory (Galmiche et al., 1999; Hooper et al., 2003, 2004; Pulford et al., 2004; Heraud et al., 2006). However, DNA-based vaccines of this sort face their own unique regulatory hurdles and are perhaps unlikely to offer significant advantages in the short term.

THE SCIENTIFIC PATHWAY TO DEVELOPMENT

The caveat noted above regarding the utility of MVA and LC16m8 as smallpox vaccines despite their historical use in West Germany and Japan, respectively—that smallpox was not endemic in either country at the time—circumscribes the major challenge faced in the development of all third-generation vaccines: the question of the extent to which confidence can be placed in a vaccine that has not been assessed against variola virus in a prospective clinical trial. Some lessons can be learned from the licensure of the second-generation vaccine ACAM2000 in the United States, but here, too, a caveat must be noted: that second-generation vaccines are expected to be effectively equivalent to the traditional vaccines insofar as they induce a take—the only established correlate of efficacy—and also have an adverse event profile indistinguishable from that of first-generation vaccines.

The treatment of severe adverse effects with vaccinia immune globulin (VIG) during and before the global eradication campaign facilitated a degree of analysis of the immune system requirements for successful vaccination. This analysis indicated that cell-mediated immune responses are necessary for successful vaccination in humans and that antibody-mediated mechanisms are less important. The latter conclusion was based on the failure of VIG to ameliorate side effects in some vaccinees with impaired cellular immunity (Freed et al., 1972). Results of subsequent studies in animals using modern techniques and reagents indicate that in fact, antibody responses play an important role in the control of orthopoxvirus infections (Belyakov et al., 2003; Edghill-Smith et al., 2005; Chaudhri et al., 2006; Heraud et al., 2006; Panchanathan et al., 2008).

The FDA has stated that in vitro neutralization studies with live variola virus would be useful in efficacy trials of third-generation vaccines (Merchlinsky, 2008; WHO, 2008). Antibody responses can be validated with recombinant antigens from variola virus produced in isolation from the virus using cloned DNA. These methods do not measure neutralizing antibodies against variola, but offer a means of comparison with the response to the homologous antigen from another orthopoxvirus, such as vaccinia virus, both as antigen produced from recombinant DNA and as part of the whole vaccinia virus. Analogous approaches may be taken to analyze cellular immune responses. Therefore, it is possible to determine whether the response to variola virus is at least similar to the response to another orthopoxvirus(es) and to correlate this response with the ability of the candidate vaccine to prevent disease induced by the test virus in an appropriate animal model. The variety of orthopoxvirus challenge models, including those that cause high levels of mortality with different pathological profiles (e.g., monkeypox in macaques, ectromelia in mice, vaccinia in mice), allows them to be to be combined in this type of approach to support the expectation that a vaccine that protects against mortality in all of these models will at the very least modify the course of disease in smallpox and increase the probability of survival.

The ability of the nonreplicating and live-attenuated virus vaccines to induce de novo production of virus proteins within host cells is an important feature shared with first- and second-generation vaccines. Nevertheless, the alterations that confer the dramatically improved safety profiles of these third-generation vaccines may plausibly have both direct and indirect effects on efficacy. The inability of protein-based subunit vaccines to direct de novo protein synthesis in the vaccinee constitutes a major departure from the first- and second-generation vaccines. Consequently, notwithstanding the efficacy of a number of third-generation vaccine approaches in animal models using nonvariola orthopoxviruses, a degree of doubt remains with regard to their potential efficacy against variola virus.

The ability to dissect the immune response induced by a vaccine does facilitate the establishment of immune correlates of protection, as has been done with, for example, vaccines against HB virus, and this can generate the necessary confidence that a vaccine is effective at either a population or individual level (Roome et al., 1993). However, concerns remain when immune profiles cannot be directly correlated with efficacy by means of prospective human trials involving the disease agent for which the vaccine is developed. For an eradicated disease, such a trial could utilize the disease agent in an animal model. However, the disconcertingly accelerated disease course and extremely high challenge dose that characterize the extant lethal variola model in nonhuman primates mean this model is inadequate for the purpose of rejecting a vaccine candidate. The model as it currently stands is thus of questionable value for the development and licensure of a third-generation vaccine.

REGULATORY REQUIREMENTS

In contrast to the challenges affecting the regulatory approval of antiviral agents for smallpox (see Chapter 6), the pathway for licensure of new vaccines is more straightforward. Although, in contrast to antivirals, the FDA has not issued formal guidance pertaining to the development and licensure of new smallpox vaccines, potentially acceptable regulatory pathways have been suggested in several publically available documents and presentations to which FDA officials have contributed.

The most pertinent event that occurred following issuance of the 1999 IOM report was the licensure of the second-generation vaccine

ACAM2000 in 2007. Approval of ACAM2000 was based primarily on clinical noninferiority in comparison with the first-generation vaccine Dryvax, with take rates, plaque reduction neutralization (PRNT) antibody responses, and acute safety parameters found to be similar in the two study groups.

The development of endpoints that could lead to the approval of third-generation vaccines has proven to be more challenging, as the accepted marker of clinical efficacy—a take—is not elicited. Under these circumstances, the Animal Rule (see Chapter 1) would play an important role in assessing efficacy, ideally in comparison with a first-generation vaccine. The FDA has also indicated that for a postevent scenario, efficacy will need to be established in at least two orthopoxvirus challenge animal models (Merchlinsky, 2008) using a dosing regimen appropriate for a postevent setting, which will most likely consist of a single dose. Moreover, because a postevent setting may also include individuals who have actually been exposed to smallpox, the time required for the induction of a protective response for a third-generation vaccine will be an important consideration in the design of animal and human studies. Finally, the use of a respiratory challenge model (preferably a nonhuman primate) should be considered, since this would be the most likely route of human exposure.

Although a path to licensure can be envisaged, the concerns raised in the previous section suggest that replacement of first- and second-generation vaccines with third-generation vaccines that do not produce lesions at the site of inoculation may be inadvisable for those segments of the population that have no contraindications for a traditional smallpox vaccine. There are nevertheless clear concerns for those segments of the population that have such contraindications. The path to licensure described above may be appropriate for less reactogenic third-generation vaccines developed specifically for these individuals, providing tangible benefits associated with, at minimum, modification of the course of disease and increased probability of survival.

For the protection of populations and individuals with contraindications, the challenge is not simply to protect against smallpox, but also to protect against adverse events associated with first- and second-generation vaccines, including contact transmission of vaccinia. It appears unlikely that a third-generation vaccine incapable of protecting these individuals against progressive vaccinia, severe generalized vaccinia, or EV would have utility against smallpox in such cases. Thus there is considerable scientific merit in focusing the development of third-generation vaccines on the prevention of adverse events associated with first- and second-generation vaccines rather than on the prevention of smallpox.

NEED FOR LIVE VARIOLA VIRUS

Although no modern prospective clinical trial of first-generation smallpox vaccines has examined protection from smallpox, experiments were conducted in the late eighteenth and early nineteenth centuries in which people were vaccinated and subsequently challenged by variolation with material taken from a smallpox patient, in an approach that would clearly be unacceptable by modern standards. The first of these were the original experiments of Edward Jenner, in which a child, James Phipps, was inoculated with cowpox by Jenner and subsequently challenged by variolation; Jenner undertook variolation challenges on two additional vaccinated children. In 1800, an American physician, Benjamin Waterhouse of Harvard University, vaccinated his son and six members of his household and subsequently arranged to have them challenged by variolation. In 1803, 17,000 vaccinations were performed in Germany; more than 8,000 of the vaccinees were subsequently challenged by variolation (Dixon, 1962). The ability to test the efficacy of vaccination by variolation challenges would necessarily have been lost in many communities as the incidence of smallpox, and thus the supply of variolation material, declined. The true efficacy of the first-generation vaccines was established through the experience of physicians and vaccinators and the success of the global eradication campaign, but there is little or no surviving evidence of evaluation in what could be considered a controlled clinical trial.

Perhaps one of the most striking advances resulting from recent work on replacement smallpox vaccines is the number of animal models that have been developed and are ready for use to examine efficacy (see Chapter 4). The basis for the success of the traditional vaccinia-based vaccine is its very close relatedness to variola. Similar levels of relatedness are apparent among all old-world orthopoxviruses, and this means they all induce a degree of protective immunity to the other members of the genus. Thus, vaccinia is able to induce immunity to smallpox and monkeypox in man, to monkeypox in monkeys, to mousepox (ectromelia) in mice, and to rabbitpox in rabbits, to name but a few. This has allowed new candidate vaccines to be extensively benchmarked against the first-generation vaccines even though no animal model using variola itself is suitable for vaccine studies.

The current status of animal models, most of which are suitable for development to Good Laboratory Practices (GLP) standards, combined with the existence of an acceptable surrogate for clinical efficacy in humans (i.e., take rates), obviates the need to use live variola virus to achieve licensure of second-generation vaccines or third-generation live-attenuated vaccines that can replicate intradermally and produce a lesion at the site of inoculation. Although the FDA has thus far indicated that licensure of nonreplicating vaccinia-based vaccines (e.g., MVA) or other third-generation vaccines for use in the general population will not necessarily require animal challenge models using live variola virus, such models, along with evidence of appropriate humoral and cellular immune responses against live variola virus in humans, would provide far more convincing evidence of efficacy. However, the development and use of such vaccines under Emergency Use Authorization (EUA) (see Chapter 1) may be justified on the basis of less stringent evidence of efficacy and may not require the use of live variola virus. For example, challenge studies based on monkeypox in nonhuman primates or other surrogate viruses, as well as neutralizing antibody and cellular immune responses in humans that are shown to be comparable to those elicited by first- or second-generation vaccines, could provide sufficient confidence for these vaccines to be used to prevent smallpox. In addition, evidence of the clinical efficacy of such a vaccine against human monkeypox disease would support such use.

While the charge to this committee was to consider variola virus, one cannot overlook the fact that the orthopoxvirus of greatest current public health concern is monkeypox. Monkeypox is endemic in central Africa and causes a severe, acute human disease that is very similar to smallpox and results in significant mortality (Hutin et al., 2001; Lederman et al., 2007; Rimoin et al., 2007). Although third-generation vaccines remain of interest for the control of potential smallpox outbreaks, their development may be more appropriately directed at the control of human monkeypox in areas where a significant proportion of the population may have medical contraindications for first- and second-generation vaccines, but are at significant risk of monkeypox virus infection.

REFERENCES

Antonie, G., F. Scheiflinger, F. Dorner, and F. G. Falkner. 1998. The complete genomic sequence of the modified vaccinia Ankara strain: comparison with other orthopoxviruses. *Virology* 244:365–396.

Artenstein, A. W., and J. D. Grabenstein. 2008. Smallpox vaccines for biodefense: Need and feasibility. *Expert Review of Vaccines* 7(8):1225–1237.

Belyakov, I. M., P. Earl, A. Dzutsev, V. A. Kuznetsov, M. Lemon, L. S. Wyatt, J. T. Snyder, J. D. Ahlers, G. Franchini, B. Moss, and J. A. Berzofsky. 2003. Shared modes of protection against poxvirus infection by attenuated and conventional smallpox vaccine viruses. *Proceedings of the National Academy of Sciences of the United States of America* 100:9458–9463.

Belyakov, I. M., D. Isakov, Q. Zhu, A. Dzutsev, D. Klinman, and J. A. Berzofsky. 2006. Enhancement of CD8+ T cell immunity in the lung by CpG oligodeoxynucleotides increases protective efficacy of a modified vaccinia Ankara vaccine against lethal poxvirus infection even in a CD4-deficient host. *Journal of Immunology* 177:6336–6343.

Blanchard, T. J., A. Alcami, P. Andrea, and G. L. Smith. 1998. Modified vaccinia virus Ankara undergoes limited replication in human cells and lacks several immunomodulatory proteins: Implications for use as a human vaccine. *Journal of General Virology* 79:1159–1167.

Bonilla-Guerrero, R., and G. A. Poland. 2003. Smallpox vaccines: Current and future. *The Journal of Laboratory and Clinical Medicine* 142(4):252–257.

Casey, C. G., J. K. Iskander, M. H. Roper, E. E. Mast, X. J. Wen, T. J. Torok, L. E. Chapman, D. L. Swerdlow, J. Morgan, J. D. Heffelfinger, C. Vitek, S. E. Reef, L. M. Hasbrouck, I. Damon, L. Neff, C. Vellozzi, M. McCauley, R. A. Strikas, and G. Mootrey. 2005. Adverse events associated with smallpox vaccination in the United States, January–October 2003. *Journal of the American Medical Association* 294(21):2734–2743.

Chapman, L. E., G. T. Mootrey, and L. J. Neff. 2008. Introduction: Vaccination against smallpox in the posteradication era. *Clinical Infectious Diseases* 46(Suppl. 3):S153–S156.

Chaudhri, G., V. Panchanathan, H. Bluethmann, and G. Karupiah. 2006. Obligatory requirement for antibody in recovery from a primary poxvirus infection. *Journal of Virology* 80:6339–6344.

Collier, L. H. 1955. The development of a stable smallpox vaccine. *Journal of Hygiene* 53:76–101.

Collier, L. H. 1980. Appropriate technology in the development of freeze-dried smallpox vaccine. *WHO Chronicle* 34:178–179.

Coulibaly, S., P. Bruhl, J. Mayrhofer, K. Schmid, M. Gerencer, and F. G. Falkner. 2005. The nonreplicating smallpox candidate vaccines defective vaccinia Lister (dVV-L) and modified vaccinia Ankara (MVA) elicit robust long-term protection. *Virology* 341:91–101.

Damon, I. K., W. B. Davidson, C. M. Hughes, V. A. Olson, S. K. Smith, R. C. Holman, S. E. Frey, F. Newman, R. B. Belshe, L. Yan, and K.S. Karem. 2009. Evaluation of smallpox vaccines using variola neutralization. *Journal of General Virology* 00(Pt 8):1962–1966.

Deng, L., P. Dai, T. Parikh, H. Cao, V. Bhoj, Q. Sun, Z. Chen, T. Merghoub, A. Houghton, and S. Shuman. 2008. Vaccinia virus subverts a mitochondrial antiviral signaling protein-dependent innate immune response in keratinocytes through its double-stranded RNA binding protein, E3. *Journal of Virology* 82(21):10735–10746.

Dixon, C. W. 1962. *Smallpox* (Chapter 12). London: J. & A. Churchill, Ltd.

Drillien, R., D. Spehner, and D. Hanau. 2004. Modified vaccinia virus Ankara induces moderate activation of human dendritic cells. *Journal of General Virology* 85:2167–2175.

Earl, P. L., J. L. Americo, L. S. Wyatt, L. A. Eller, J. C. Whitbeck, G. H. Cohen, R. J. Eisenberg, C. J. Hartmann, D. L. Jackson, D. A. Kulesh, M. J. Martinez, D. M. Miller, E. M. Mucker, J. D. Shamblin, S. H. Zwiers, J. W. Huggins, P. B. Jahrling, and B. Moss. 2004. Immunogenicity of a highly attenuated MVA smallpox vaccine and protection against monkeypox. *Nature* 428:182–185.

Earl, P. L., J. L. Americo, L. S. Wyatt, O. Espenshade, J. Bassler, K. Gong, S. Lin, E. Peters, L. Rhodes Jr., Y. E. Spano, P. M. Silvera, and B. Moss. 2008. Rapid protection in a monkeypox model by a single injection of a replication-deficient vaccinia virus. *Proceedings of the National Academy of Sciences of the United States of America* 105:10889–10894.

Edghill-Smith, Y., H. Golding, J. Manischewitz, L. R. King, D. Scott, M. Bray, A. Nalca, J. W. Hooper, C. A. Whitehouse, J. E. Schmitz, K. A. Reimann, and G. Franchini. 2005. Smallpox vaccine-induced antibodies are necessary and sufficient for protection against monkeypox virus. *Nature Medicine* 11:740–747.

Fauci, A. S. 2002. Bioterrorism: Defining a research agenda. *Food and Drug Law Journal* 57(3):413–421.

Fenner, F., D. A. Henderson, I. Arita, Z. Jezek, and I. D. Ladnyi. 1988. *Smallpox and its eradication*. Geneva, Switzerland: WHO.

Fogg, C., S. Lustig, J. C. Whitbeck, R. J. Eisenberg, G. H. Cohen, and B. Moss. 2004. Protective immunity to vaccinia virus induced by vaccination with multiple recombinant outer membrane proteins of intracellular and extracellular virions. *Journal of Virology* 78(19):10230–10237.

Freed, E. R., R. J. Dumar, and M. R. Escobar. 1972. Vaccinia necrosum and its relationship to impaired immunologic responsiveness. *American Journal of Medicine* 52:411–420.

Frey, S. E., F. K. Newman, J. S. Kennedy, F. Ennis, G. Abate, D. F. Hoft, and T. P. Monath. 2009. Comparison of the safety and immunogenicity of ACAM1000, ACAM2000 and Dryvax® in healthy vaccinia-naive adults. *Vaccine* 27(10):1637–1644.

Galmiche, M. C., J. Goenaga, R. Wittek, L. Rindisbacher. 1999. Neutralizing and protective antibodies directed against vaccinia virus envelope antigens. *Virology* 254(1):71–80.

Greenberg, R. N., and J. S. Kennedy. 2008. ACAM2000: A newly licensed cell culture-based live vaccinia smallpox vaccine. *Expert Opinion on Investigational Drugs* 17(4):555–564.

Heraud, J. M., Y. Edghill-Smith, V. Ayala, I. Kalisz, J. Parrino, V. S. Kalyanaraman, J. Manischewitz, L. R. King, A. Hryniewicz, C. J. Trindade, M. Hassett, W. P. Tsai, D. Venzon, A. Nalca, M. Vaccari, P. Silvera, M. Bray, B. S. Graham, H. Golding, J. W. Hooper, and G. Franchini. 2006. Subunit recombinant vaccine protects against monkeypox. *Journal of Immunology* 177:2552–2564.

Hochstein-Mintzel, V., T. Hanichen, H. C. Huber, and H. Stickl. 1975. An attenuated strain of vaccinia virus (MVA). Successful intramuscular immunization against vaccinia and variola (author's translation). *Zentralbl Bakteriol* 230:283–297.

Hooper, J. W., D. M. Custer, and E. Thompson. 2003. Four-gene-combination DNA vaccine protects mice against a lethal vaccinia virus challenge and elicits appropriate antibody responses in nonhuman primates. *Virology* 306:181–195.

Hooper, J. W., E. Thompson, C. Wilhelmsen, M. Zimmerman, M. A. Ichou, S. E. Steffen, C. S. Schmaljohn, A. L. Schmaljohn, and P. B. Jahrling. 2004. Smallpox DNA vaccine protects nonhuman primates against lethal monkeypox. *Journal of Virology* 78:4433–4443.

Horii, K. A., S. D. Simon, D. Y. Liu, and V. Sharma. 2007. Atopic dermatitis in children in the United States, 1997–2004: Visit trends, patient and provider characteristics, and prescribing patterns. *Pediatrics* 120(3):e527–e534.

Howell, M. D., J. F. Jones, K. O. Kisich, J. E. Streib, R. L. Gallo, and D. Y. Leung. 2004. Selective killing of vaccinia virus by LL-37: implications for eczema vaccinatum. *Journal of Immunology* 172(3):1763–1767.

Howell, M. D., R. L. Gallo, M. Boguniewicz, J. F. Jones, C. Wong, J. E. Streib, and D. Y. Leung. 2006. Cytokine milieu of atopic dermatitis skin subverts the innate immune response to vaccinia virus. *Immunity* 24(3):341–348.

Hutin, Y. J., R. J. Williams, P. Malfait, R. Pebody, V. N. Loparev, S. L. Ropp, M. Rodriguez, J. C. Knight, F. K. Tshioko, A. S. Khan, M. V. Szczeniowski, and J. J. Esposito. 2001. Outbreak of human monkeypox, Democratic Republic of Congo, 1996 to 1997. *Emerging Infectious Diseases* 7(3):434–438.

IOM (Institute of Medicine). 2003a. *Review of the Centers for Disease Control and Prevention's Smallpox Vaccination Program Implementation: Letter Report #1*. Washington, DC: The National Academies Press.

IOM. 2003b. *Review of the Centers for Disease Control and Prevention's Smallpox Vaccination Program Implementation: Letter Report #2*. Washington, DC: The National Academies Press.

IOM. 2003c. *Review of the Centers for Disease Control and Prevention's Smallpox Vaccination Program Implementation: Letter Report #3*. Washington, DC: The National Academies Press.

IOM. 2003d. *Review of the Centers for Disease Control and Prevention's Smallpox Vaccination Program Implementation: Letter Report #4*. Washington, DC: The National Academies Press.

IOM. 2003e. *Review of the Centers for Disease Control and Prevention's Smallpox Vaccination Program Implementation: Letter Report #5*. Washington, DC: The National Academies Press.

IOM. 2004. *Review of the Centers for Disease Control and Prevention's Smallpox Vaccination Program Implementation: Letter Report #6*. Washington, DC: The National Academies Press.

IOM. 2005. *Public Health in an Age of Terrorism*. Washington, DC: The National Academies Press.

Kemper, A. R., M. M. Davis, and G. L. Freed. 2002. Expected adverse events in a mass smallpox vaccination campaign. *Effective Clinical Practice: ECP* 5:84–90.

Kenner, J., F. Cameron, C. Empig, D. V. Jobes, and M. Gurwith. 2006. LC16m8: An attenuated smallpox vaccine. *Vaccine* 24(47–48):7009–7022.

Kidokoro, M., M. Tashiro, and H. Shida. 2005. Genetically stable and fully effective smallpox vaccine strain constructed from highly attenuated vaccinia LC16m8. *Proceedings of the National Academy of Sciences of the United States of America* 102(11):4152–4157.

Lane, J. M., F. L. Ruben, J. M. Neff, and J. D. Millar. 1969. Complications of smallpox vaccination, 1968. *New England Journal of Medicine* 281:1201–1208.

Lederman, E. R., M. G. Reynolds, K. Karem, Z. Braden, L. A. Learned-Orozco, D. Wassa-Wassa, O. Moundeli, C. Hughes, J. Harvey, R. Regnery, J. V. Mombouli, and I. K. Damon. 2007. Prevalence of antibodies against orthopoxviruses among residents of Likouala region, Republic of Congo: Evidence for monkeypox virus exposure. *American Journal of Tropical Medicine and Hygiene* 77(6):1150–1156.

Lewis, F. S., S. A. Norton, R. D. Bradshaw, J. Lapa, and J. D. Grabenstein. 2006. Analysis of cases reported as generalized vaccinia during the US military smallpox vaccination program, December 2002 to December 2004. *Journal of the American Academy of Dermatology* 55(1):23–31.

Liu, L., X. Zhan, R. C. Fuhlbrigge, V. Pena-Cruz, J. Lieberman, and T. S. Kupper. 2005. Vaccinia virus induces strong immunoregulatory cytokine production in healthy human epidermal keratinocytes: a novel strategy for immune evasion. *Journal of Virology* 79 (12):7363–7370.

Maksyutov, R. A., I. N. Babkina, A. E. Nesterov, and S. N. Shchelkunov. 2006. Development of candidate DNA vaccine against human orthopoxvirus infections. *Biotechnolgiya* 4:23–30.

Mahnel, H. 1985. Vaccination against mouse pox. *Tierarztl Prax* 13:403–407.

Mahnel, H., and A. Mayr. 1994. Experiences with immunization against orthopox viruses of humans and animals using vaccine strain MVA. *Berl Munch Tierarztl Wochenschr* 107:253–256.

Mayr, A., H. Stickl, H. K. Muller, K. Danner, and H. Singer. 1978. The smallpox vaccination strain MVA: Marker, genetic structure, experience gained with the parenteral vaccination and behavior in organisms with a debilitated defence mechanism (author's translation). *Zentralbl Bakteriol [B]* 167:375–390.

McCurdy, L. H., B. D. Larkin, J. E. Martin, and B. S. Graham, 2004. Modified vaccinia Ankara: Potential as an alternative smallpox vaccine. *Clinical Infectious Diseases* 38:1749–1753.

Merchlinsky, M. 2008. Regulatory issues with licensing new generation smallpox vaccines CBER/FDA October 3, 2008. Presentation to the Committee. Washington, DC.

Meseda, C. A., A. D. Garcia, A. Kumar, A. E. Mayer, J. Manischewitz, L. R. King, H. Golding, M. Merchlinsky, and J. P. Weir. 2005. Enhanced immunogenicity and protective effect conferred by vaccination with combinations of modified vaccinia virus Ankara and licensed smallpox vaccine Dryvax in a mouse model. *Virology* 339:164–175.

Meyer, H., G. Sutter, and A. Mayr. 1991. Mapping of deletions in the genome of the highly attenuated vaccinia virus MVA and their influence on virulence. *Journal of General Virology* 72:1031–1038.

Morgan, J., M. H. Roper, L. Sperling, R. A. Schieber, J. D. Heffelfinger, C. G. Casey, J. W. Miller, S. Santibanez, B. Herwaldt, P. Hightower, P. L. Moro, B. F. Hibbs, N. H. Levine, L. E. Chapman, J. Iskander, J. M. Lane, M. Wharton, G. T. Mootrey, and D. L. Swerdlow. 2008. Myocarditis, pericarditis, and dilated cardiomyopathy after smallpox vaccination among civilians in the United States, January–October 2003. *Clinical Infectious Diseases* 46(Suppl. 3):S242–S250.

Neff, J., J. Modlin, G. S. Birkhead, G. Poland, R. M. Robertson, K. Sepkowitz, C. Yancy, P. Gardner, G. C. Gray, T. Maurer, J. Siegel, F. A. Guerra, T. Berger, W. D. Flanders, and R. Shope. 2008. Monitoring the safety of a smallpox vaccination program in the United States: Report of the joint Smallpox Vaccine Safety Working Group of the advisory committee on immunization practices and the Armed Forces Epidemiological Board. *Clinical Infectious Diseases* 46(Suppl. 3):S258–S270.

Panchanathan, V., G. Chaudhri, and G. Karupiah. 2008. Correlates of protective immunity in poxvirus infection: Where does antibody stand? *Immunology and Cell Biology* 86:80–86.

Paoletti, E., J. Tartaglia, and J. Taylor. 1994. Safe and effective poxvirus vectors—NYVAC and ALVAC. *Developments in Biological Standardization* 82:65–69.

Parrino, J., and B. S. Graham. 2006. Smallpox vaccines: Past, present, and future. *Journal of Allergy and Clinical Immunology* 118(6):1320–1326.

Parrino, J., L. H. McCurdy, B. D. Larkin, I. J. Gordon, S. E. Rucker, M. E. Enama, R. A. Koup, M. Roederer, R. T. Bailer, Z. Moodie, L. Gu, L. Yan, B. S. Graham, and VRC 201/203 Study Team. 2007. Safety, immunogenicity and efficacy of modified vaccinia Ankara (MVA) against Dryvax challenge in vaccinia-naïve and vaccinia-immune individuals. *Vaccine* 25(8):1513–1525.

Phelps, A. L., A. J. Gates, M. Hillier, L. Eastaugh, and D. O. Ulaeto. 2007. Comparative efficacy of modified vaccinia Ankara (MVA) as a potential replacement smallpox vaccine. *Vaccine* 25:34–42.

Pliasunov, I. V., A. N. Sergeev, A. A. Sergeev, V. A. Petrishchenko, L. N. Shishkina, V. V. Generalov, A. S. Safatov, L. S. Sandakhchiev, V. V. Udut, S. A. Mel'nikov, and V. N. Podkuĭko. 2006. Clinical trials of oral recombinant bivaccine against variola and hepatitis B during double vaccination. *Voprosy Birusologii* 51(2):31–35.

Poland, G. A., J. D. Grabenstein, and J. M. Neff. 2005. The U.S. smallpox vaccination program: a review of a large modern era smallpox vaccination implementation program. *Vaccine* 23(17–18):2078–2081.

Pulford, D. J., A. Gates, S. H. Bridge, J. H. Robinson, and D. Ulaeto. 2004. Differential efficacy of vaccinia virus envelope proteins administered by DNA immunisation in protection of BALB/c mice from a lethal intranasal poxvirus challenge. *Vaccine* 22:3358–3366.

Rimoin, A. W., N. Kisalu, B. Kebela-Ilunga, T. Mukaba, L. L. Wright, P. Formenty, N. D. Wolfe, R. L. Shongo, F. Tshioko, E. Okitolonda, J. J. Muyembe, R. W. Ryder, and H. Meyer. 2007. Endemic human monkeypox, Democratic Republic of Congo, 2001–2004. *Emerging Infectious Diseases* 13(6):934–937.

Roome, A. J., S. J. Walsh, M. L. Cartter, and J. L. Hadler. 1993. Hepatitis B vaccine responsiveness in Connecticut public safety personnel. *Journal of American Medical Association* 270 (24):2931–2934.

Saijo, M., Y. Ami, Y. Suzaki, N. Nagata, N. Iwata, H. Hasegawa, M. Ogata, S. Fukushi, T. Mizutani, T. Sata, T. Kurata, I. Kurane, S. Morikawa. 2006. LC16m8, a highly attenuated vaccinia virus vaccine lacking expression of the membrane protein B5R, protects monkeys from monkeypox. *Journal of Virology* 80(11):5179–5188.

Samuelsson, C., J. Hausmann, H. Lauterbach, M. Schmidt, S. Akira, H. Wagner, P. Chaplin, M. Suter, M. O'Keeffe, and H. Hochrein. 2008. Survival of lethal poxvirus infection in mice depends on TLR9, and therapeutic vaccination provides protection. *Journal of Clinical Investigation* 118:1776–1784.

Seiler, N., H. Taylor, and R. Faden. 2003. Legal and ethical considerations in government compensation plans: A case study of smallpox immunization. *Indiana Health Law Review* 1(1): 3–27.

Sergeev, A. A., A. N. Sergeev, V. A. Petrishchenko, L. N. Shishkina, G. V. Kochneva, V. A. Zhukov, N. K. Evtin, O. V. P'iankov, L. S. Sandakhchiev, L. A. Akinfeeva, V. V. Udut, E. Sherstoboev, E. D. Gol'dberg, S. A. Mel'nikov, V. N. Podkuĭko, and V. A. Maksimov. 2004. Reactogenicity, safety and immunogenicity of a recombinant bivaccine against smallpox and hepatitis B in limited clinical trials. *Voprosy Birusologii* 49(5):22–26.

Slifka, M. K. 2005. The future of smallpox vaccination: Is MVA the key? *Medical Immunology (London, England)* 4:2.

Stittelaar, K. J., T. Kuiken, R. L. de Swart, G. van Amerongen, H. W. Vos, H. G. Niesters, P. van Schalkwijk, T. van der Kwast, L. W. Wyatt, B. Moss, and A. D. Osterhaus. 2001. Safety of modified vaccinia virus Ankara (MVA) in immune-suppressed macaques. *Vaccine* 19:3700–3709.

Stittelaar, K. J., G. van Amerongen, I. Kondova, T. Kuiken, R. F. van Lavieren, F. H. Pistoor, H. G. Niesters, G. van Doornum, B. A. van der Zeijst, L. Mateo, P. J. Chaplin, and A. D. Osterhaus. 2005. Modified vaccinia virus Ankara protects macaques against respiratory challenge with monkeypox virus. *Journal of Virology* 79:7845–7851.

Strikas, R. A., L. J. Neff, L. Rotz, J. Cono, D. Knutson, J. Henderson, and W. A. Orenstein. 2008. U.S. Civilian Smallpox Preparedness and Response Program, 2003. *Clinical Infectious Diseases* 46(Suppl. 3):S157–S167.

Swerdlow, D. L., M. H. Roper, J. Morgan, R. A. Schieber, L. S. Sperling, M. M. Sniadack, L. Neff, J. W. Miller, C. R. Curtis, M. E. Marin, J. Iskander, P. Moro, P. Hightower, N. H. Levine, M. McCauley, J. Heffelfinger, I. Damon, T. J. Torok, M. Wharton, E. E. Mast, and G. T. Mootrey. 2008. Ischemic cardiac events during the Department of Health and Human Services Smallpox Vaccination Program, 2003. *Clinical Infectious Diseases* 46(Suppl. 3):S234–S241.

Tartaglia, J., M. E. Perkus, J. Taylor, E. K. Norton, J. C. Audonnet, W. I. Cox, S. W. Davis, J. van der Hoeven, B. Meignier, and M. Riviere. 1992. NYVAC: A highly attenuated strain of vaccinia virus. *Virology* 188:217–232.

Thomas, T. N., S. Reef, L. Neff, M. M. Sniadack, and G. T. Mootrey. 2008. A review of the smallpox vaccine adverse events active surveillance system. *Clinical Infectious Diseases* 46(Suppl. 3):S212–S220.

Vollmar, J., N. Arndtz, K. M.Eckl, T. Thomsen, B. Petzold, L. Mateo, B. Schlereth, A. Handley, L. King, V. Hülsemann, M. Tzatzaris, K. Merkl, N. Wulff, and P. Chaplin. 2006. Safety and immunogenicity of IMVAMUNE, a promising candidate as a third generation smallpox vaccine. *Vaccine* 24:2065–2070.

Vora, S., I. Damon, V. Fulginiti, S. G. Weber, M. Kahana, S. L. Stein, S. I. Gerber, S. Garcia-Houchins, E. Lederman, D. Hruby, L. Collins, D. Scott, K. Thompson, J. V. Barson, R. Regnery, C. Hughes, R. S. Daum, Y. Li, H. Zhao, S. Smith, Z. Braden, K. Karem, V. Olson, W. Davidson, G. Trindade, T. Bolken, R. Jordan, D. Tien, and J. Marcinak. 2008. Severe eczema vaccinatum in a household contact of a smallpox vaccinee. *Clinical Infectious Diseases* 46(10):1555–1561.

Waibler, Z., M. Anzaghe, H. Ludwig, S. Akira, S. Weiss, G. Sutter, and U. Kalinke. 2007. Modified vaccinia virus Ankara induces Toll-like receptor-independent type I interferon responses. *Journal of Virology* 81(22):12102–12110.

WHO (World Health Organization). 2008. WHO Advisory Committee on Variola Virus Research Report of the Tenth Meeting, November 19–20. Geneva, Switzerland: WHO. http://www.who.int/csr/resources/publications/WHO_HSE_EPR_2008_9/en/index.html (accessed March 23, 2009).

Wilson, S. J. 2005. Factors affecting implementation of the U.S. smallpox vaccination program. *Public Health Reports* 120:3–5.

Wollenberg, A., and R. Engler. 2004. Smallpox, vaccination and adverse reactions to smallpox vaccine. *Current Opinion in Allergy and Clinical Immunology* 4(4):271–275.

Wyatt, L. S., P. L. Earl, L. A. Eller, and B. Moss. 2004. Highly attenuated smallpox vaccine protects mice with and without immune deficiencies against pathogenic vaccinia virus challenge. *Proceedings of the National Academy of Sciences of the United States of America* 101:4590–4595.

Yamaguchi, M., M. Kimura, and M. Hirayama. 1975. Report of the National Smallpox Vaccination Research Committee: study of side effects, complications and their treatments. *Clinical Virology* 3:269–278 (in Japanese).

8

Methods for Detection and Diagnosis

In the pre-eradication era, smallpox was usually diagnosed by its distinct clinical characteristics, particularly the vesicular–pustular rash (see Chapter 2), in the context of a cluster of probable cases with an epidemiologic link. Material from lesions could be analyzed by electron microscopy in reference laboratories, providing morphologic identification of the characteristic brick-shaped virions (Biel and Gelderblom, 1999; see also Chapter 3), but did not distinguish variola from other poxviruses. Recovery of infectious virus from infected persons using tissue culture methods was feasible but was seldom used.

Despite the eradication of smallpox, the need remains for robust and safe methods of detection of variola virus and diagnosis of the disease. Diseases caused by poxviruses that can infect the human host, such as monkeypox, continue to circulate and may be confused with smallpox, necessitating precise methods for rapid differential diagnosis. Disseminated vaccinia might also be misdiagnosed as smallpox, although a history of recent vaccination or contact with a recently vaccinated person would usually be obtained. Finally, the classification of smallpox as a category A agent with the potential for aerosolization and broad distribution within the environment requires new approaches to sensitive and specific detection of the virus in nonclinical specimens.

The application of contemporary viral diagnostic tools, such as polymerase chain reaction (PCR) methods, to smallpox diagnosis has received attention because rapid and accurate identification of index cases would be essential for optimal containment of initial spread in a largely unimmunized population in the event of an unintended or intentional release of the virus.

These methods may allow diagnosis in respiratory secretions during the 12–14 day incubation period, which would be quite valuable for controlling transmission. Additionally, recent and forthcoming advances in genomic science mean that new approaches for identification of variola virus in clinical or environmental samples can be developed that involve detecting the presence of genomic DNA or viral proteins. Maximizing the specificity of such tests will require knowledge of the genetic variability of related poxviruses, the background against which variola must be distinguished to maximize the sensitivity of the test, and the variability of variola and viral proteins and their subdomains that are unique to variola. It will also be important to develop new diagnostics that can be used to detect the virus in different types of patient specimens (e.g., lesion material, secretions, organ tissues) and environmental samples (e.g., air, surfaces, fomites). Developing environmental detection and diagnostic methods that do not require the isolation of infectious virus in tissue culture is important because of the risk of human exposure during preparation of specimens to be tested in the laboratory. Such advances in detection and diagnosis would facilitate forensic investigations to determine the source of variola virus in the event of an intentional release.

This chapter reviews the current status of methods to detect variola virus and diagnose smallpox, relevant regulatory requirements, and the need for live variola virus to achieve advances in the development of detection and diagnostic capabilities.

CURRENT STATUS OF DETECTION AND DIAGNOSTIC METHODS

The 1999 committee offered the following conclusion related to detection and diagnosis:

> **If further development of procedures for the environmental detection of variola virus or for diagnostic purposes were to be pursued, more extensive knowledge of the genome variability, predicted protein sequences, virion surface structure, and functionality of variola virus from widely dispersed geographic sources would be needed.**

Since 1999, substantial work has been done on the development of new techniques for the detection of variola virus and diagnosis of smallpox and for the differentiation of variola virus from other orthopoxviruses that infect humans (e.g., vaccinia, monkeypox, cowpox). Most of these assays have been based on nucleic acid detection by PCR, and some have been validated using clinical samples. Some experience has been reported with the use of multiplex PCR to detect variola and differentiate it from other poxviruses or unrelated viruses in laboratory-created specimens containing mixed genomic fragments. Limited experience exists with direct detection of variola virus in stored patient specimens or in specimens from nonhuman primates. Relatively little has been done to create assays that detect variola virus proteins or to refine serologic approaches to smallpox diagnosis. The capacity to carry out seroepidemiologic surveillance with rapid high-throughput serologic assays for variola virus-specific IgG antibodies would be valuable to characterize the extent of the spread of the virus in an outbreak setting, and serologic assays for variola virus-specific IgM antibodies would be useful to document recent infection in individuals who were asymptomatic when tested (see Appendix).

Polymerase Chain Reaction

PCR enables highly sensitive detection of viral nucleic acids to very low copy numbers. PCR products can be sequenced to provide detailed genetic information about the pathogen, and PCR can be performed as a quantitative or multiplex assay in which the specimen is tested for multiple pathogen sequences at the same time. Several different regions of the variola virus genome have been used to design primers that either detect all orthopoxviruses of interest or are specific for individual poxviruses. Real-time PCR for the hemagglutinin gene (J7R) of variola virus was sensitive and specific when tested on variola virus samples from cell culture and infected tissues that contained both viral and cellular DNA (Ibrahim et al., 2003; Aitichou et al., 2008). This assay was evaluated with genomic DNA from 48 different isolates of variola virus and 25 other poxviruses. Specificity for variola detection was greater than 96 percent; the majority of these samples were derived from virus-infected cell cultures and variola virus-infected tissues. This poxvirus assay was applied successfully to the diagnosis of smallpox from fixed human tissue from one fatal case (Schoepp et al., 2004), even though specimens were obtained and stored under conditions not designed to protect DNA integrity. The assay has been expanded to include other variola virus genes (B9R and B10R) using prepared samples, detecting 12–25 genome copies (Kulesh et al., 2004). It has been adapted for use with dried reagents and for multiplexing with probes for other orthopoxviruses (Aitichou et al., 2008). The hemagglutinin gene has also been used to design primers for detecting all orthopoxviruses for use with a probe that can distinguish variola from other poxviruses by melting curve analysis, and tested on plasmid DNA (Espy et al., 2002) and on tissue and blood spiked with poxvirus DNA (Putkuri et al., 2009).

The CrmB (cytokine response modifier B) gene has also served as the target for amplifying orthopoxvirus DNA using consensus primers. Viral (genomic) amplicons may differ in size, but variola and other orthopoxviruses can also be differentiated from each other by analysis of restriction fragment length polymorphism (RFLP) (Loparev et al., 2001). This assay was validated on eight strains of variola virus. In a similar assay, TaqMan probes were designed to be specific for all orthopoxviruses or for variola virus and validated with poxvirus panels and plasmid DNA from the European Network for Imported Viral Diseases (Fedele et al., 2006). A multiplex PCR that distinguished orthopoxviruses from herpesviruses used primers from the CrmB gene for poxvirus identification and RFLP of the PCR product to differentiate one orthopoxvirus from another. This test was developed and validated using plasmid DNA from only a single strain of variola virus (Sias et al., 2007). A real-time PCR assay that combines variola virus-specific and panorthopoxvirus primers targeted to the gene for a 14 Kd protein (A30L) has been developed and validated on genomic DNA from 12 strains of variola virus; variola was differentiated from cowpox, vaccinia, monkeypox, and camelpox viruses (Scaramozzino et al., 2007).

A multiplex real-time PCR assay has been developed that includes individual primers specific for variola (B11R–B12R), vaccinia, monkeypox, and cowpox viruses, plus primers common for all orthopoxviruses, and results in amplicons of different sizes. This assay was validated on DNA from virus grown in culture and on scabs from smallpox skin lesions (Shchelkunov et al., 2005). Another multiplex method targets the 14kD fusion protein (A27L) for amplification from all orthopoxviruses and differentiates variola from other orthopoxviruses by melting curve analysis (Olson et al., 2004). This assay was validated on 14 variola virus samples from tissue culture and from skin lesions in the VECTOR repository and detected four variola genome copies. Multiplex PCR has also been performed using consensus and variola virus-specific primers based on known single nucleotide polymorphisms (SNPs) in A13L and A36R genes that are different in variola and other poxviruses; these SNPs were identified in PCR products from 43 variola strains but none of 50 other orthopoxviruses (Pulford et al., 2004). These variola virus isolates had been collected over 40 years from diverse geographic locations.

A number of PCR assays have been developed and tested for detection and differentiation of variola virus using only plasmid DNA. The genes analyzed include hemagglutinin, RNA polymerase (rpo18), early transcription factor VETF, and small membrane protein p8 (A13L). For each, melting curve analysis was used to distinguish variola from other orthopoxviruses (Nitsche et al., 2004; Panning et al., 2004). PCR has also been combined with immobilization of synthetic oligonucleotides corresponding to variola and other poxvirus genes on nylon membranes to allow direct visualization of products that hybridize to specific oligonucleotides as a simplification, but a PCR apparatus is still required (Fitzgibbon and Sagripanti, 2006).

In Situ Hybridization

In situ hybridization was used to examine sections of tissue specimens for the presence of variola virus DNA in skin lesion biopsies from two South American smallpox cases. Specific molecular probes differentiated skin cells containing variola from those caused by herpesviruses in formalin-fixed tissue sections that showed no distinguishable differences by standard histopathology methods (Nuovo et al., 2003).

Gene Chip Analysis

Oligonucleotides specific for orthopoxviruses can be immobilized and used to detect interaction with DNA extracted from samples suspected of containing a poxvirus. Specific hybridization can be detected by fluorescent probes (Lapa et al., 2002; Laassri et al., 2003; Ryabinin et al., 2006) or use of electrochemical sensors (Komarova et al., 2005). Chips have been designed using one or two individual variola virus genes (CrmB, Lapa et al., 2002, and Komarova et al., 2005; C23L/B29R, Laassri et al., 2003; C23L/B29R + B19R, Ryabinin et al., 2006) and the complete genomes of multiple strains as resequencing tiling arrays (Sulaiman et al., 2007). These assays can distinguish variola virus from other poxviruses and from herpesviruses. The resequencing array was tested on amplified DNA from 14 strains of variola virus and can also identify other human orthopoxviruses (Sulaiman et al., 2008). This technology can be used for rapid identification of a particular variola genome by comparison with known genomes in sequencing databases. A variation on this approach is the development of primers that span the orthopoxvirus genome followed by RFLP, which is then used to distinguish one orthopoxvirus from another. This assay was validated on genomic DNA from two strains of variola virus and on monkeypox, camelpox, cowpox, tanapox, ectromelia, and vaccinia viruses (Li et al., 2007). These whole-genome approaches would be useful to identify variola genomes that had been altered intentionally.

Protein-Based Methods

Little work has been done to develop direct protein detection methods for variola. At present, these methods depend on developing antibody reagents that bind specifically to variola proteins that are distinct from those made by other orthopoxviruses. Utilizing ELISAs, Ulaeto and colleagues (2002) have begun to characterize the reactivity of 23 strains/isolates of variola virus, both $\gamma^{®}$-ray inactivated and viable (under BSL-4 conditions), with a panel of monoclonal antibodies and polyclonal antisera, raised against either vaccinia or variola virus preparations. Polyclonal antibody reagents displayed more uniform detection of variola virus strains than was obtained with monoclonal antibodies (Ulaeto et al., 2002). One monoclonal antibody has been described that is specific for variola virus and can be used to distinguish variola from other poxviruses (Damon, 2006). However, monoclonal antibodies detect a single epitope in a single viral protein, and most are conformation dependent. Specificity for geographically unrelated variola isolates would depend on defining a fully conserved and stable epitope or using a mix of monoclonals that would recognize epitopes in several unique variola virus proteins having no homologues or differing substantially from the related proteins in the other poxviruses.

Even when well-characterized reagents are available, designing antigen detection methods that demonstrate the presence of viral proteins in patient materials has been challenging for many human pathogens. Most successes are achieved when the clinical material is a cutaneous lesion specimen, which would be the case for variola at the symptomatic stage of infection. In one example of a poxvirus detection method applied to respiratory secretions, a biosensor technique using cyan-5 dye labeled antivaccinia antibody was used to detect vaccina proteins in human throat swab specimens that had been spiked with vaccinia virus from tissue culture (Donaldson et al., 2004). One would expect such approaches to be feasible for variola detection, but their development currently depends on generating panels of antibodies that are highly specific for variola proteins. Pilot experiments were conducted in which ELISAs were used to detect monkeypox virus during the recent outbreaks in Africa and variola virus in specimens from nonhuman primates (Karem et al., 2007). Nevertheless, although inhibitors may be encountered, nucleotide detection methods are generally preferred for viral detection because nucleotides can be extracted from patient materials and concentrated for PCR testing, whereas similar processes to enhance sensitivity are difficult for protein detection in respiratory secretions or other clinical specimens that would be available from patients in the pre-eruptive phase of smallpox. Proteomics methods may emerge that can identify a specific sequence of amino acid residues by direct analysis of a sample using mass spectroscopy or other methods that do not require antibody reagents, but these tools are not yet applicable for clinical use.

With the exception of measuring antibody titers by plaque reduction neutralization assay, serologic assays for IgG and IgM antibodies to variola and other poxviruses are also protein-based detection techniques. ELISA methods detect antibodies in serum samples through their binding to immobilized viral antigens. The development of such an assay for detecting variola virus IgG and IgM antibodies is feasible, but specificity requires the identification of unique proteins that do not elicit cross-reactive antibodies as a result of exposure to other poxviruses, such as by vaccination with vaccinia. It is anticipated that most variola infections would be symptomatic;

however, a panpoxvirus serologic assay could be useful for assessing the extent of asymptomatic infection in a population not previously vaccinated should the need arise.

Currently, VECTOR is developing next-generation test kits to detect orthopoxviral protein markers. These immunodiagnostic tests will rely on hybridoma technology and technology for producing recombinant antibodies to major neutralizing and protective antigens of variola virus and those of other orthopoxviruses pathogenic for humans (Russian federation Patent #2281327; Razumov et al., 2004, 2005; personal communication, Ilya Drozdov, WHOCC for Orthopoxvirus Diagnosis and Repository for Variola Virus Strains and DNA, March 27, 2009). In parallel, VECTOR is working to develop and improve species-specific diagnostics for viruses such as variola, monkeypox, and cowpox based on multiplex PCR, real-time PCR, and microchip technology (Lapa et al., 2002; Laassri et al., 2003; Ryabinin et al., 2006; personal communication, Ilya Drozdov, WHOCC for Orthopoxvirus Diagnosis and Repository for Variola Virus Strains and DNA, March 27, 2009).

Detection in the Environment

The technical capacity for environmental detection of variola virus would be important in the event of an intentional release. Widespread distribution of the virus could be achieved because poxviruses are stable in aerosol form and can be lyophilized. The molecular methods for variola virus detection that have been developed since 1999 use PCR and in situ hybridization assays that have proven valuable for the clinical detection of many viral pathogens in patient specimens, and a few of these methods have been validated using archived tissues from variola cases. PCR-based methods are also useful for detecting viruses in environmental samples, including air samples, water, and soil, as well as in swabs taken from potentially contaminated surfaces. These methods could be applied to the identification of variola virus in such specimens with certain modifications in the way the materials are prepared for testing. For example, it would be necessary to take into account the inhibitory effects of detergents and other materials on PCR sensitivity, as shown in experiments with vaccinia virus (Kurth et al., 2008). The specificity of PCR for variola virus detection should be preserved, but sensitivity in such samples is difficult to predict.

Ideally, tools for detecting the presence of variola virus in the environment would need to be rapid, portable, and easily deployable. Because poxvirus genome detection methods require relatively complex equipment and reagents, it would be necessary at present to bring materials suspected of containing variola virus to a laboratory facility. A more practical variation of the method for field use would be the use of dried reagents in a dual-probe real-time PCR assay for detection of variola or other orthopoxviruses (Aitichou et al., 2008). Even if PCR or ELISA methods were used that could differentiate variola from other poxviruses in environmental samples, their sensitivity in field testing would need to be established. Criteria for specificity might need to be lowered to ensure that a positive sample was not missed under field conditions, with the assumption that all specimens would need to be retested and results validated in a reference laboratory. This gap may be addressed by the development of tools such as direct electrochemical DNA sensors that can identify nucleotide sequences without the need for PCR amplification and secondary analysis of the products by RFLP or sequencing (Komarova et al., 2005). Nanotechnology-based tools may be developed that can discriminate viruses based on their particle size and other properties; if so, it would be necessary to have at least inactivated variola particles to assess their sensitivity for environmental detection.

REGULATORY REQUIREMENTS

Currently available in vitro diagnostic devices (IVDs) for the detection and diagnosis of variola infection are limited to research assays developed by DOD, CDC, and academic laboratories. In the United States, licensure of IVDs for various infectious agents, including variola, is regulated primarily by the FDA's Center for Devices and Radiological Health (CDRH), which assesses benefits and risks according to the IVD's analytical and clinical performance. Medical devices, including IVDs, are categorized as Class I, II, or III according to risk criteria and requirements listed in 21 Code of Federal Regulations (CFR) 800. Whereas most Class I devices are exempt from premarket notification, most Class II devices do require such notification [510(k)], and most Class III devices require premarket approval (PMA), including submission of clinical data to support marketing claims. The potential classification of IVDs for variola virus detection has not been established, although it appears likely, given the critical importance of accurate detection methods, that premarket notification including both general and special controls (Class II designation) would be required. A new section (513(f)(2)) of the Food, Drug and Cosmetics Act as amended by the FDA Modernization Act of 1997 includes a provision whereby a sponsor can request a so-called "de novo" classification that may not require premarket approval, but the sponsor would have to demonstrate that the device would pose very little or no risk of harm, especially for diagnosing suspected human cases. Finally, the use of a new IVD for variola virus detection may also be approved via Emergency Use Authorization (see Chapter 1).

NEED FOR LIVE VARIOLA VIRUS

The identification and characterization of a series of variola virus-specific genetic markers has paved the way for sensitive and specific multiplex nucleic acid methods, and further progress on refining these approaches should not require live virus. Methods that detect viral proteins have been pursued to a lesser extent but could also be expanded without the need for live virus. Although not essential, better characterization of the sensitivity and specificity of both nucleic acid and protein methods for variola virus detection in relevant samples could be achieved by additional testing of tissues from nonhuman primates infected with the virus. Preservation of tissues for this purpose should be included in antiviral, vaccine, or pathogenesis studies done in animals infected with variola. Since methods developed using only variola proteins could prove inadequate for their detection in clinical materials from infected individuals, archived clinical specimens could be tested to confirm the sensitivity and specificity of such tests, if possible. Further work on protein-based detection would benefit particularly from access to proteins made in variola virus-infected cells instead of proteins made using expression vectors to ensure the reliability of the test and to standardize reagents.

Environmental detection methods have seen little progress, but further research in this area would use "mocked-up" specimens, so use of the live virus would not be necessary. High-throughput assays, including serologic methods to identify recently infected individuals, would be needed to test large numbers of samples in a possible outbreak situation. However, the development of most new methods would not require live virus as this research could build on work with other validated methods and be scaled up. Some future approaches that might prove valuable, such as those that detect viral particles, could require access to variola virions made in culture cells for their validation.

One caveat related to variola detection and smallpox diagnosis is that genomic sequencing of enough geographically diverse isolates is necessary to ensure that PCR tests have adequate specificity. PCR and sequencing of the amplicons would be the first step in a forensic analysis of the source of a variola isolate should a reintroduction of the virus occur, and would also be accomplished most effectively if background information were available on the complete genome sequence of as many variola isolates as possible. It is expected that use of the live virus would not be necessary for this purpose, assuming that sufficient DNA is still available in stored specimens in the U.S. and Russian stocks.

Finally, it is not yet clear whether the FDA will require the use of live variola virus in the evaluation of new diagnostic methods.

REFERENCES

Aitichou, M., S. Saleh, P. Kyusung, J. Huggins, M. O'Guinn, P. Jahrling, and S. Ibrahim. 2008. Dual-probe real-time PCR assay for detection of variola or other orthopoxviruses with dried reagents. *Journal of Virological Methods* 153(2):190–195.

Biel, S. S., and H. R. Gelderblom. 1999. Diagnostic electron microscopy is still a timely and rewarding method. *Journal of Clinical Virology* 13:105–119.

Damon, I. K. 2006. Poxviruses. In *Fields' virology*, 5th edition, edited by B. N. Fields, D. M. Knipe, P. M. Howley, and D. E. Griffin. Philadelphia, PA: Lippincott Williams & Wilkins. Pp. 2947–2976.

Donaldson, K. A., M. F. Kramer, and D. V. Lim. 2004. A rapid detection method for Vaccinia virus, the surrogate for smallpox virus. *Biosens Bioelectron* 20(2):322–327.

Espy, M. J., I. F. Cockerill, R. F. Meyer, M. D. Bowen, G. A. Poland, T. L. Hadfield, and T. F. Smith. 2002. Detection of smallpox virus DNA by LightCycler PCR. *Journal of Clinical Microbiology* 40(6):1985–1988.

Fedele, C. G., A. Negredo, F. Molero, M. P. Sanchez-Seco, and A. Tenorio. 2006. Use of internally controlled real-time genome amplification for detection of variola virus and other orthopoxviruses infecting humans. *Journal of Clinical Microbiology* 44(12):4464–4470.

Fitzgibbon, J. E., and J. L. Sagripanti. 2006. Simultaneous identification of orthopoxviruses and alphaviruses by oligonucleotide macroarray with special emphasis on detection of variola and Venezuelan equine encephalitis viruses. *Journal of Virological Methods* 131(2):160–167.

Ibrahim, S. M., D. A. Kulesh, S. S. Saleh, I. K. Damon, J. J. Esposito, A. L. Schmaljohn, and P. B. Jahrling. 2003. Real-time PCR assay to detect smallpox virus. *Journal of Clinical Microbiology* 41(8):3835–3839.

Karem, K.L., M. Reynolds, C. Hughes, Z. Braden, P. Nigam, S. Crotty, J. Glidewell, R. Ahmed, R. Amara, and I. K. Damon. 2007. Monkeypox-induced immunity and failure of childhood smallpox vaccination to provide complete protection. *Clinical & Vaccine Immunology: CVI* 14 (10):1318–1327.

Komarova, E., M. Aldissi, and A. Bogomolova. 2005. Direct electrochemical sensor for fast reagent-free DNA detection. *Biosens Bioelectron* 21(1):182–189.

Kulesh, D. A., R. O. Baker, B. M. Loveless, D. Norwood, S. H. Zwiers, E. Mucker, C. Hartmann, R. Herrera, D. Miller, D. Christensen, L. P. Wasieloski, Jr., J. Huggins, and P. B. Jahrling. 2004. Smallpox and pan-orthopox virus detection by real-time 3'-minor groove binder TaqMan assays on the roche LightCycler and the Cepheid smart Cycler platforms. *Journal of Clinical Microbiology* 42(2):601–609.

Kurth, A., J. Achenbach, L. Miller, I. M. Mackay, G. Pauli, and A. Nitsche. 2008. Orthopoxvirus detection in environmental specimens during suspected bioterror attacks: Inhibitory influences of common household products. *Applied and Environmental Microbiology* 74(1):32–37.

Laassri, M., V. Chizhikov, M. Mikheev, S. Shchelkunov, and K. Chumakov. 2003. Detection and discrimination of orthopoxviruses using microarrays of immobilized oligonucleotides. *Journal of Virological Methods* 112(1-2):67–78.

Lapa, S., M. Mikheev, S. Shchelkunov, V. Mikhailovich, A. Sobolev, V. Blinov, I. Babkin, A. Guskov, E. Sokunova, A. Zasedatelev, L. Sandakhchiev, and A. Mirzabekov. 2002. Species-level identification of orthopoxviruses with an oligonucleotide microchip. *Journal of Clinical Microbiology* 40(3):753–757.

Li, Y., D. S. Carroll, S. N. Gardner, M. C. Walsh, E. A. Vitalis, and I. K. Damon. 2007. On the origin of smallpox: Correlating variola phylogenics with historical smallpox records. *Proceedings of the National Academy of Sciences of the United States of America* 104:15787–15792.

Loparev, V. N., R. F. Massung, J. J. Esposito, and H. Meyer. 2001. Detection and differentiation of Old World orthopoxviruses: Restriction fragment length polymorphism of the crmB gene region. *Journal of Clinical Microbiology* 39(1):94–100.

Nitsche, A., H. Ellerbrok, and G. Pauli. 2004. Detection of orthopoxvirus DNA by real-time PCR and identification of variola virus DNA by melting analysis. *Journal of Clinical Microbiology* 42(3):1207–1213.

Nuovo, G. J., J. A. Plaza, and C. Magro. 2003. Rapid diagnosis of smallpox infection and differentiation from its mimics. *Diagnostic Molecular Pathology* 12(2):103–107.

Olson, V. A., T. Laue, M. T. Laker, I. V. Babkin, C. Drosten, S. N. Shchelkunov, M. Niedrig, I. K. Damon, and H. Meyer. 2004. Real-time PCR system for detection of orthopoxviruses and simultaneous identification of smallpox virus. *Journal of Clinical Microbiology* 42(5):1940–1946.

Panning, M., M. Asper, S. Kramme, H. Schmitz, and C. Drosten. 2004. Rapid detection and differentiation of human pathogenic orthopox viruses by a fluorescence resonance energy transfer real-time PCR assay. *Clinical Chemistry* 50(4):702–708.

Pulford, D. J., A. Gates, S. H. Bridge, J. H. Robinson, and D. Ulaeto. 2004. Differential efficacy of vaccinia virus envelope proteins administered by DNA immunisation in protection of BALB/c mice from a lethal intranasal poxvirus challenge. *Vaccine* 22:3358–3366.

Putkuri, N., H. Piiparinen, A. Vaheri, and O. Vapalahti. 2009. Detection of human orthopoxvirus infections and differentiation of smallpox virus with real-time PCR. *Journal of Medical Virology* 81(1):146–152.

Razumov, I. A., M. A. Vasil'eva, O. A. Serova, L. V. Artem'eva, N. I. Bormotov, E. F. Belanov, G. V. Kochneva, E. E. Konovalov, and V. B. Loktev. 2004. A study of neutralizing activity and cross-reactivity of monoclonal antibodies to ectromelia virus with orthopoxviruses pathogenic for man. *Vestn Ross Akad Med Nauk* (8):19–22. Article in Russian.

Razumov, I. A., Gileva, I. P., M. A. Vasil'eva, T. S. Nepomniashchikh, M. N. Mishina, E. F. Belanov, G. V. Kochneva, E. E. Konovalov, S. N. Shchelkunov, and V. B. Loktev. 2005. Neutralizing monoclonal antibodies cross-react with fusion proteins encoded by 129L of the ectromelia virus and A30L of the variola virus. *Molecular Biology* 39(6):918–925.

Ryabinin, V. A., L. A. Shundrin, E. B. Kostina, M. Laassri, V. Chizhikov, S. N. Shchelkunov, K. Chumakov, and A. N. Sinyakov. 2006. Microarray assay for detection and discrimination of Orthopoxvirus species. *Journal of Medical Virology* 78(10):1325–1340.

Scaramozzino, N., A. Ferrier-Rembert, A. L. Favier, C. Rothlisberger, S. Richard, J. M. Crance, H. Meyer, and D. Garin. 2007. Real-time PCR to identify variola virus or other human pathogenic orthopox viruses. *Clinical Chemistry* 53(4):606–613.

Schoepp, R. J., M. D. Morin, M. J. Martinez, D. A. Kulesh, L. Hensley, T. W. Geisbert, D. R. Brady, and P. B. Jahrling. 2004. Detection and identification of variola virus in fixed human tissue after prolonged archival storage. *Laboratory Investigation: A Journal of Technical Methods and Pathology* 84(1):41–48.

Shchelkunov, S. N., E. V. Gavrilova, and I. V. Babkin. 2005. Multiplex PCR detection and species differentiation of orthopoxviruses pathogenic to humans. *Molecular & Cellular Probes* 19(1):1–8.

Sias, C., F. Carletti, M. R. Capobianchi, D. Travaglini, R. Chiappini, D. Horejsh, and A. Di Caro. 2007. Rapid differential diagnosis of orthopoxviruses and herpesviruses based upon multiplex real-time PCR. *Infez Med* 15(1):47–55.

Sulaiman, I. M., K. Tang, J. Osborne, S. Sammons, and R. M. Wohlhueter. 2007. GeneChip resequencing of the smallpox virus genome can identify novel strains: A biodefense application. *Journal of Clinical Microbiology* 45(2):358–363.

Sulaiman, I. M., S. A. Sammons, and R. M. Wohlhueter. 2008. Smallpox virus resequencing GeneChips can also rapidly ascertain species status for some zoonotic non-variola orthopoxviruses. *Journal of Clinical Microbiology* 46(4):1507–1509.

Ulaeto, D., et al., Poster presentation. XIV International Poxvirus &. Iridovirus Workshop, Lake Placid, New York, September 19–25, 2002.

9

Discovery Research

The combination of recent technological advances in molecular biology, genomics, and computational biology, coupled with the intimate relationship between variola virus and the human immune system in particular, creates unusual opportunities for scientific discovery. The guiding rationale for work with variola is the possibility of obtaining novel insights that would lead to new smallpox prevention strategies, diagnostic approaches, and therapeutic interventions. Given that variola exclusively infects humans under natural conditions and has adapted to specifically modulate the human immune system, much could be learned about human biology from studies with this virus. As variola proteins that dampen or manipulate a particular immune response are identified, these viral proteins, or portions thereof, become candidate novel therapeutics for autoimmune or inflammatory diseases in which the host response is aberrant or overactive.

The 1999 IOM committee offered two conclusions related to discovery research:

- **Live or replication-defective variola virus would be needed if studies of variola pathogenesis were to be undertaken to provide information about the response of the human immune system.**
- **Variola virus proteins have potential as reagents in studies of human immunology. Live variola virus would be needed for this purpose only until sufficient variola isolates had been cloned and sequenced.**

The 1999 IOM committee acknowledged that variola virus plays a unique role in contributing to understanding of the human immune system. Research in this area could yield further information about human-specific reagents with therapeutic or immunomodulatory potential.

This chapter examines opportunities for discovery research involving variola virus in three areas: the potential to gain new insights into the pathogenesis of smallpox through the capabilities offered by systems biology, understanding of the subversion and modulation of human immune responses, and the possibilities for development of novel variola-based therapeutics. The final section addresses the need for live variola virus to conduct this work.

SYSTEMS BIOLOGY AND SMALLPOX PATHOGENESIS

While some progress has been made since 1999 toward elucidating the pathogenesis of smallpox and characterizing viral immunomodulatory activities, much more remains to be learned. The synthesis of molecular biology, genomics, and computational biology, or "systems biology," offers promising approaches for understanding smallpox pathogenesis, human immunology, and other aspects of host defense and for identifying novel therapeutic targets and strategies.

Systems biology refers to the study of the behavior of complex biological organization and processes in terms of the molecular constituents (Kirschner, 2005). It is made possible by the availability of broad-based, genome-wide, high-throughput approaches for measuring the abundance and localization of DNA, RNA, and protein and their interactions within an entire biological system. Although in its early days, this discipline offers the promise of revealing rules and features that can lead to predictions about the vulnerabilities and control points of a cell or an organism. For instance, this approach could be used to examine the interaction between variola and an infected target cell, or the broader interactions between variola and an infected host as in studies by Rubins and colleagues (2004) (see below), but with a more complete set of measurements of RNA and protein. The use of systems biology techniques in a more comprehensive and integrated fashion represents a largely untapped resource for learning more about the variola life cycle and the interactions between variola virus and its host.

Limited studies of pathogenesis have been performed with variola virus in nonhuman primates. Jahrling and colleagues (2004) describe a model of lethal disease in cynomolgus macaques with features of late, severe smallpox, achieved using high intravenous doses of variola virus (the Harper and India 7124 strains). DNA microarrays were performed in peripheral blood cells from these infected animals to examine host gene expression patterns (Rubins et al., 2004) (see Chapter 4). From these data, groups of genes were identified, as well as coregulated biological processes associated with these genes and their products. Some of these processes, such as "cell proliferation," had not been emphasized previously as a prominent feature of variola infection of primates. Although these studies revealed important information regarding smallpox pathogenesis—such as the prominence of an interferon-associated response; unusual suppression of the NFkB response system; and the possible importance of other biological processes, such as cell proliferation—they were limited in a number of ways. First, the nonhuman primate model was inadequate for studying early aspects of smallpox as it occurred naturally in humans. In addition, the kinds of measurements performed in these studies were limited and did not include the newly discovered and critical noncoding RNAs of primates, or genome-wide patterns of protein expression, or the interactions of proteins and nucleic acids, all of which can now be quantified using high-throughput genome-wide technologies. Furthermore, the responses of different individual cell types have not yet been explored, even though it is clear that distinct biology is found in different cells. With today's improved high-level containment research facilities and more powerful research technologies, a great deal more might be learned about variola–host interactions, with relevance to the development of smallpox therapeutics.

Poxviruses replicate in the cytoplasm of susceptible host cells and contain regulatory sequence elements that are virus specific (Moss, 1996). An accurate functional analysis of poxvirus proteins may require expression systems that replicate the posttranslational modifications found in naturally infected cells and hosts, and may not be possible with typical protein expression systems (e.g., bacteria, yeast, or insect cells). Moreover, some viral proteins may have multiple, unrelated functions or may function primarily as a complex with other viral or host cell proteins, and thus may be biologically inert if expressed in the wrong cell type or in the absence of a productive infection. Bearing this in mind, it is possible that some variola proteins will require analysis in the context of live infection of human cells or through coexpression experiments with a number of other viral protein partners.

SUBVERSION AND MODULATION OF HUMAN IMMUNE RESPONSES

At the time of the 1999 IOM report, it was known that poxviruses encode the largest number of putative immunomodulatory proteins of any group of mammalian viruses (Barry and McFadden, 1997). As of this writing, only five putative immunomodulatory proteins from variola virus have been characterized: D12/SPICE (smallpox inhibitor of complement enzymes) (Liszewski et al., 2008), G3R/CKBP-II (variola virus high-affinity secreted chemokine-binding protein type II) (Smith et al., 1997), B9R (an IFN-g inhibitor) (Seregin et al., 1996), G2R (a tumor necrosis factor inhibitor) (Alejo et al., 2006), and D5L (an IL-18-binding protein) (Esteban et al., 2004). The findings from this work, although limited, suggest that variola proteins have potent biological activity and may have special value in blunting human immune responses. Some of these findings are summarized below to illustrate the possible rewards of future work on these and other variola proteins, which could also yield insights into the mechanisms of variola pathogenesis.

Variola CrmB encodes a tumor necrosis factor receptor (TNFR) homologue that acts as a soluble decoy of TNFR, as well as a chemokine inhibitor through its C terminal domain—the smallpox virus-encoded chemokine receptor (SECRET) domain (Alejo et al., 2006). This is the first example of a dual function for a poxvirus decoy molecule. The SECRET domain was subsequently identified in another variola TNFR homologue (CrmD) and three other orthopoxvirus-encoded secreted proteins (Alejo et al., 2006).

Both variola and ectromelia virus encode soluble decoys that inhibit the activity of IL-18, an important proinflammatory cytokine (Esteban et al., 2004). Using surface plasmon resonance, it has been shown that both proteins have higher affinity for murine than for human IL-18, which is similar to human IL-18BP and an ortholog encoded by molluscum contagiosum virus (Xiang and Moss, 1999). Variola IL-18-binding protein (IL-18BP) also binds to glycoaminoglycans, whereas the ectomelia ortholog does not (Esteban et al., 2004). The 2.0-Å resolution crystal structure of a binary complex human IL-18 and ectromelia IL-18BP was recently solved (Krumm et al., 2008), and reveals significant conformational changes at the binding interface. The residues of ectromelia IL-18BP at the interface are conserved in both human IL-18BP and viral homologues.

Although functional analysis of related immunomodulatory proteins from other orthopoxviruses can provide insight into the activities of their variola-encoded counterparts, this approach may not always provide an accurate understanding of the virulence factors of variola. For instance, direct comparison of the variola-encoded complement inhibitor SPICE with similar evasion proteins encoded by vaccinia (VCP) and monkeypox (MOPICE) revealed that VCP and MOPICE were approximately 100-fold less efficient than SPICE (Liszewski et al., 2006). This work suggests that studies involving viral gene products from even closely related orthopoxviruses will not necessarily provide the same information that would be attained by directly examining variola virulence proteins and immune evasion proteins. A relatively restricted number of variola proteins have been studied in detail, and although this work represents an important step forward, much remains to be learned about these and other variola gene products.

The 1999 IOM report suggested that it would be possible to study variola protein products in isolation once a number of virus isolates had been sequenced. However, given the regulatory hurdles facing the use of variola-specific gene sequences and the restrictions associated with the use of live variola virus, most studies on poxvirus immune evasion have been performed with related orthopoxviruses instead of variola. More than a dozen predicted immunomodulatory proteins encoded by variola major have yet to be fully tested and characterized (McFadden, 2004).

Recent studies conducted with related orthopoxviruses, such as cowpox and monkeypox, have revealed previously unrealized immune evasion/subversion mechanisms that may be relevant to smallpox pathogenesis. For instance, cowpox expresses proteins that downregulate MHC Class I molecules on the infected cell surface (Byun et al., 2007; Dasgupta et al., 2007). By reducing MHC Class I expression, the virus is able to evade recognition by cytolytic $CD8^+$ T cells in a manner similar to the evasion strategies employed by many herpesviruses. Monkeypox has developed an even more intriguing strategy of host immune system manipulation by triggering a nonresponsive state in either $CD4^+$ or $CD8^+$ T cells that come into direct contact with monkeypox-infected monocytes (Hammarlund et al., 2008). These studies were performed by infecting primary human peripheral blood monocytes and measuring cytokine production by poxvirus-specific T cells using intracellular cytokine staining analysis. It is not known whether variola expresses similar or possibly an even more extensive battery of immunomodulatory genes that could directly block human T cell recognition and/or antiviral function in similar in vitro experiments. Also unknown is whether different strains of variola major and variola minor differ with respect to their ability to evade host T cell responses. Variations in the expression of various immunomodulatory proteins could explain the dramatic differences in pathogenesis and mortality rates that are associated with these two forms of smallpox, as well as with different strains of variola major.

In addition to evading host T cell responses, poxviruses are known to subvert antiviral innate immune responses, including the attenuation of type I interferon, proinflammatory cytokines, and chemokine production. Upon infection, viral nucleic acids can be sensed through a variety of pathways by host immune cells to trigger an immune response. Toll-like receptor (TLR)3, TLR7, and TLR9 are endosomal TLRs that recognize dsRNA, ssRNA, and viral DNA. RIG-I and MDA-5 are cytosolic RNA sensors (Kawai and Akira, 2008). Signaling through these pathways leads to type I interferon production and NF-κB activation. AIM2 is a recently identified cytosolic DNA sensor that may link DNA virus infection to inflammasome activation (Hornung et al., 2009). How poxviruses are sensed in various immune cells has just begun to be understood. Ectromelia virus (the causative agent of mousepox) activates pDCs through TLR9, and mice lacking TLR9 are more susceptible to ectromelia infection (Samuelsson et al., 2008). Infection of murine keratinocytes with vaccinia virus containing a deletion of the immunomodulatory gene E3L triggers a vigorous innate immune response that is dependent on cytosolic RNA sensing pathway and transcription factor IRF3 (Deng et al., 2008). Human macrophages sense myxoma virus and produce type I interferon and TNF-α that is dependent on RIG-I and IRF3 (Wang et al., 2008). Overall, then, poxviruses can be sensed by different pathways in a variety of immune cells to trigger an antiviral response, but there is a relative dearth of information about variola virus and its specific interactions with the human innate immune system.

NOVEL VARIOLA-BASED THERAPEUTICS

Genome sequences from 45 strains of variola currently provide a set of diverse variola-specific proteins and variants that might be expressed and screened for biological activities of interest. These proteins could themselves serve as immunomodulatory agents or might provide leads for the development of related molecules. Additional variola genome sequences from as yet uncharacterized variola isolates might be expected to yield new sequence variants and expand this set of potential novel biologicals. It should be noted that specialized expression systems may be necessary for critical specific posttranslational modifications of these proteins. Moreover, some viral proteins may have multiple, unrelated functions or may function primarily in a complex or in concert with other viral or host cell proteins, and thus may fail to demonstrate the relevant phenotype if expressed in the wrong cell type or in the absence of a productive variola virus infection.

Many viral immunomodulatory proteins act with high specificity against a particular immune function or pathway and do so at very low doses (femtomolar to nanomolar), making these proteins potentially feasible for use as therapeutics to treat diseases of overactive immune function or inflammation (McFadden and Murphy, 2000; Shisler and Moss, 2001; Johnston and McFadden, 2003; Seet et al., 2003). The field of virogenomics (Fruh et al., 2001; Kellam, 2001; DeFilippis et al., 2003) is emerging as a means of future drug discovery and will continue to flourish as knowledge and understanding of host–pathogen interactions increases.

NEED FOR LIVE VARIOLA VIRUS

Comparative studies of variola major and variola minor in primary human cells have not been performed. The differences in virulence between the two or among different strains of variola major may lie in these interactions. Because variola was eradicated prior to the marked advances in modern cellular immunology and molecular biology techniques that have since occurred, understanding of human immune responses to variola infection remains very limited. Live variola virus would be required to perform these comparative in vitro studies. Live virus would also be needed for the use of systems biology approaches in an improved nonhuman primate model with the goal of identifying novel therapeutic targets.

With more than 40 strains of variola now having been sequenced, there is ample opportunity to study specific variola proteins, and thereby advance understanding of host–pathogen interactions and develop potential new therapeutic drugs. Live virus would be useful for initiating some of these studies but would not be required for most of this research.

REFERENCES

Alejo, A., M. B. Ruiz-Argüello, Y. Ho, V. P. Smith, M. Saraiva, and A. Alcami. 2006. A chemokine-binding domain in the tumor necrosis factor receptor from variola (smallpox) virus. *Proceedings of the National Academy of Sciences of the United States of America* 103:5995–6000.

Barry, M., and G. McFadden. 1997. Virokines and viroceptors. In *Cytokines in health and disease*, edited by D. G. Remick and J. S. Friedland. New York: Marcel Dekker.

Byun, M., X. Wang, M. Pak, T. H. Hansen, and W. M. Yokoyama. 2007. Cowpox virus exploits the endoplasmic reticulum retention pathway to inhibit MHC class I transport to the cell surface. *Cell Host Microbe* 2(5):306.

Dasgupta, A., E. Hammarlund, M. K. Slifka, and K. Früh. 2007. Cowpox virus evades CTL recognition and inhibits the intracellular transport of MHC class I molecules. *Journal of Immunology* 178:1654.

DeFilippis, V., C. Raggo, A. Moses, and K. Früh. 2003. Functional genomics in virology and antiviral drug discovery. *Trends in Biotechnology* 21(10):452–457.

Deng, L., P. Dai, T. Parikh, H. Cao, V. Bhjoy, Q. Sun, Z. J. Chen, T. Merghoub, A. Houghton, and S. Shuman. 2008. Vaccinia virus subverts a mitochondrial antiviral signaling protein-dependent innate immune response in keratinocytes through its dsRNA binding protein E3. *Journal of Virology* 82:10735–10746.

Esteban, D. J., A. A. Nuara, and M. L. Buller. 2004. Interleukin-18 and glycosaminoglycan binding by a protein encoded by variola virus. *Journal of General Virology* 85(Pt. 5):1291–1299.

Fruh, K., K. Simmen, B. G. Luukkonen, Y. C. Bell, and P. Ghazal. 2001. Virogenomics: A novel approach to antiviral drug discovery. *Drug Discovery Today* 6 (12):621–627.

Hammarlund, E., A. Dasgupta, C. Pinilla, P. Norori, K. Früh, and M. K. Slifka. 2008. Monkeypox virus evades antiviral CD4+ and CD8+ T cell responses by suppressing cognate T cell activation. *Proceedings of the National Academy of Sciences of the United States of America* 105:14567.

Hornung, V., A. Ablasser, M. Charrel-Dennis, F. Bauernfeind, G. Horvath, D. R. Caffrey, E. Latz, and K. A. Fitzgerald. 2009. AIM2 recognizes cytosolic dsDNA and forms a caspase-1-activating inflammasome with ASC. *Nature* 458(7237):514–518.

Jahrling, P. B., L. E. Hensley, M. J. Martinez, J. W. LeDuc, K. H. Rubins, D. A. Relman, and J. W. Huggins. 2004. Variola virus infection of cynomolgus macaques: A model for human smallpox. *Proceedings of the National Academy of Sciences of the United States of America* 101:15196–15200.

Johnston, J. B., and G. McFadden. 2003. Poxvirus immunomodulatory strategies: Current perspectives. *Journal of Virology* 77(11):6093–6100.

Kawai, T., and S. Akira. 2008. Toll-like receptor and RIG-I-like receptor signaling. *Annals of the New York Academy of Sciences* 1143:1–20.

Kellam, P. 2001. Post-genomic virology: The impact of bioinformatics, microarrays and proteomics on investigating host and pathogen interactions. *Reviews in Medical Virology* 11(5):313–329.

Kirschner, M. W. 2005. The meaning of systems biology. *Cell* 121:503–504.

Krumm, B., X. Meng, Y. Li, Y. Xiang, and J. Deng. 2008. Structural basis for antagonism of human interleukin 18 by poxvirus interleukin 18-binding protein. *Proceedings of the National Academy of Sciences of the United States of America* 105(52):20711–20715.

Liszewski, M. K., M. K. Leung, R. Hauhart, R. M. L. Buller, P. Bertram, X. Wang, A. M. Rosengard, G. J. Kotwal, and J. P. Atkinson. 2006. Structure and regulatory profile of the monkeypox inhibitor of complement: Comparison to homologs in vaccinia and variola and evidence for dimer formation. *Journal of Immunology* 176(6):3725–3734.

Liszewski, M. K., P. Bertram, M. K. Leung, R. Hauhart, L. Zhang, and J. P. Atkinson. 2008. Smallpox inhibitor of complement enzymes (SPICE): Regulation of complement activation on cells and mechanism of its cellular attachment. *Journal of Immunology* 181:4199–4207.

McFadden, G. 2004. Smallpox: An ancient disease enters the modern era of virogenomics. *Proceedings of the National Academy of Sciences of the United States of America* 101(42):14994–14995.

McFadden, G., and P. M. Murphy. 2000. Host-related immunomodulators encoded by poxviruses and herpesviruses. *Current Opinion in Microbiology* 3(4):371–378.

Moss, B. 1996. Replication of poxviruses. In *Fields virology*, 3rd edition, edited by B. N. Fields, D. M. Knipe, and P. M. Howley. Philadelphia: Lippincott-Raven.

Rubins, K., L. E. Hensley, P. B. Jahrling, A. A. Whitney, T. W. Geisbert, J. W. Huggins, A. Owen, J. W. LeDuc, P. O. Brown, and D. A. Relman. 2004. The host response to smallpox: Analysis of the gene expression program in peripheral blood cells in a monkey model. *Proceedings of the National Academy of Sciences of the United States of America* 101:15190–15195.

Samuelsson, C., J. Hausmann, H. Lauterbach, M. Schmidt, S. Akira, H. Wagner, P. Chaplin, M. Suter, M. O'Keeffe, and H. Hochrein. 2008. Survival of lethal poxvirus infection in mice depends on TLR9, and therapeutic vaccination provides protection. *Journal of Clinical Investigation* 118:1776–1784.

Seet, B. T., J. B. Johnston, C. R. Brunetti, J. W. Barrett, H. Everett, C. Cameron, J. Sypula, S. H. Nazarian, A. Lucas, and G. McFadden. 2003. Poxviruses and immune evasion. *Annual Review of Immunology* 21:377–423.

Seregin, S. V., I. N. Babkina, A. E. Nesterov, A. N. Sinyakov, and S. N. Shchelkunov. 1996. Comparative studies of gamma-interferon receptor-like proteins of variola major and variola minor viruses. *FEBS Letters* 382(1–2):79–83.

Shisler, J. L., and B. Moss. 2001. Immunology 102 at poxvirus U: avoiding apoptosis. *Seminars in Immunology* 13(1):67–72.

Smith, G. L., T. D. Smith, P. J. Smolak, D. Friend, H. Hagen, M. Gerhart, L. Park, D. J. Pickup, D. Torrance, K. Mohler, K. Schooley, R. G. Goodwin. 1997. Poxvirus genomes encode a secreted, soluble protein that preferentially inhibits beta chemokine activity yet lacks sequence homology to known chemokine receptors. *Virology* 236:316–327.

Wang, F., X. Gao, J. W. Barrett, Q. Shao, E. Bartee, M. R. Mohamed, M. Rahman, S. Werden, T. Irvine, J. Cao, G. A. Dekaban, and G. McFadden. 2008. RIG-I mediates the co-induction of tumor necrosis factor and type I interferon elicited by myxoma virus in primary human macrophages. *PLoS Pathogens* 4(7):e1000099.

Xiang, Y., and B. Moss. 1999. IL-8 binding and inhibition of interferon gamma induction by human poxvirus-encoded proteins. *Proceedings of the National Academy of Sciences of the United States of America* 96(20):11537–11542.

10

Conclusions and Recommendations

Thirty years have passed since WHO declared smallpox eradicated. Since then, programs for universal vaccination against smallpox have ceased worldwide, yielding a growing population of immunologically naïve individuals; U.S. and international regulatory requirements for licensure of antiviral drugs and vaccines have become better defined; and technological advances in molecular biology have generated sophisticated tools for research and development, many of which have been applied to improving knowledge about variola virus. Given that an accidental or deliberate release of variola virus could have devastating results worldwide, current global public health preparedness efforts address the potential threat of a smallpox outbreak. WHO considers any confirmed case of smallpox to be a public health emergency of international concern, and the U.S. government classifies this pathogen as a select agent.

This committee was not asked to consider whether live variola virus stocks should be retained or destroyed or to address the potential for a smallpox outbreak. Nevertheless, these issues underlie global deliberations about smallpox, and the development and availability of adequate medical countermeasures against one of the most virulent and dangerous pathogens remains a strategic international goal.

Variola is a unique and highly adapted pathogen that has established a close and obligate relationship with the human species, its only natural host. While not immediately essential, research that advances understanding of the biology of the human species and its responses to life-threatening microbial challenges could be highly beneficial. Such research could provide fundamental insights into human physiology and immunology that would be relevant for biomedical research, leading to new therapies and preventive measures.

Capabilities in DNA synthesis, sequence error correction, and assembly of custom-designed long DNA molecules have grown exponentially over the past decade. It is now technically feasible to chemically synthesize and assemble a complete variola genome in the laboratory, although the subsequent steps necessary for production of intact, replication-competent virions are likely to be challenging. It is uncertain that variola virions generated from synthetic variola genomes would be virulent for humans, and if so to what degree. However, fully virulent synthetic variola virus is a distinct possibility. This disconcerting reality should be acknowledged because it has major implications for the risks associated with unregulated possession and genetic manipulation of variola virus. These advances also offer potential benefits for the future development of variola countermeasures.

In this contemporary context, some research with live variola virus remains essential for public health preparedness, some would be useful for this purpose, and some would have significant scientific merit as biomedical research without an immediate connection to preparedness. All research with live variola virus requires rigorous scientific evaluation before being undertaken, proper laboratory safeguards to protect those working with the virus and the public, and a significant investment in public health infrastructure and research capacity. Research to develop and improve diagnostics and preventive and therapeutic countermeasures against smallpox must also be undertaken with specific attention to regulatory concerns. While the scientific pathway for development of these diagnostics and countermeasures may offer a spectrum of options, from ideal to practical, the absence of human infection presents special challenges for regulatory approval. Regulatory agencies must evaluate new interventions that are of potential but unproven value for the prevention and treatment of smallpox and establish appropriate contingency protocols for their use in the event of an accidental or intentional release. These interventions may also warrant evaluation against nonvariola poxvirus infections, such as disseminated vaccinia or monkeypox disease, under conditions that make standard clinical trials difficult or impossible to accomplish.

CONCLUSIONS

This committee, like its predecessor in 1999, did not consider the risk assessment or financial resources required to undertake necessary or useful research, as these issues were beyond its scope. In addition, since decision making can be based only on information in hand, the committee recognizes that future technological advances or policy considerations based on assessment of the risk of an accidental or intentional release of variola virus could alter the scientific landscape. With these caveats, the committee offers the following conclusions, which are based on its evaluation of current scientific capabilities and are meant to address the question of whether live variola virus would be needed should the recommended research be undertaken.

Development of Therapeutics

The discovery of antiviral drugs and alternative therapeutic agents effective against smallpox and their advanced development through licensure and postlicensure is vital. Such agents are needed for the medical management of variola infection, a critical element in preparedness for a rapid response to an outbreak. Antiviral agents with good oral bioavailability that are effective for prophylaxis as well as treatment are important for containing the spread of smallpox in an immunologically naïve population. Having more than one licensed therapeutic utilizing multiple mechanisms of action is desirable because of the potential for the emergence of drug resistance and unanticipated adverse effects. Even if multiple licensed drugs were available, there would be gaps in information regarding their safety in special populations, such as children or pregnant women. If an appropriate clinical context is available, such as a monkeypox outbreak or cases of eczema vaccinatum, candidate drugs should be assessed in these groups.

The development of licensed therapeutics is a long-term effort. Over the last decade, substantial progress has been made in the development of antiviral drugs with potential efficacy against smallpox using surrogate orthopoxviruses. Live variola virus has been used to measure the activity of lead candidate drugs in vitro and in nonhuman primate models. Additional studies are needed to develop useful drugs and immunobiologics through discovery efforts aimed at identifying variola-specific targets. This undertaking will require a better understanding of variola-specific proteins and their functions in cultured cells and of how these gene products contribute to the pathogenesis of smallpox disease in suitable animal models.

The committee concludes that, for both scientific and regulatory reasons, the final developmental stages leading to licensure of smallpox therapeutics cannot occur without the use of live variola virus. Furthermore, although the regulatory environment may change, the scientific reasons will remain. Therapeutic agents need to be evaluated against a representative panel of variola strains to reduce the possibility that some strains might be naturally resistant.

Development of Vaccines

The availability and strategic deployment of an effective vaccine enabled the eradication of smallpox. Despite the occurrence of adverse reactions, enough people worldwide were vaccinated and developed immunity sufficient to interrupt transmission. Today, the majority of the world's population is unvaccinated, placing them at risk of life-threatening disease in the case of a smallpox outbreak. Should an outbreak of smallpox occur, scaling up immunization programs with the traditional vaccines could be expected to be effective again. However, vaccine safety would be of particular concern for the substantial number of immunocompromised individuals and other vulnerable populations.

Since the 1999 IOM report was issued, traditional vaccines such as Dryvax and the Lister/Elstree vaccine, which were manufactured by being grown in animals, have been augmented by the production and licensure of second-generation vaccines using modern tissue culture techniques. For first- and second-generation vaccines, successful vaccination is manifested by a "take"—formation of a lesion at the site of inoculation. This method cannot be used for evaluation of third-generation vaccines, and immunologic correlates of protection cannot be defined in the absence of circulating variola virus. Evidence that would support likely efficacy can be obtained only in animal model studies using variola virus. It should be emphasized that populations for whom the use of first- and second-generation vaccines would be contraindicated would need to rely on safer third-generation vaccines in the event of an outbreak. Some consideration should be given to methods that could accelerate the pathway to licensure (or at least approved use) in these populations.

The committee concludes that the current development and licensure pathway for first- and second-generation vaccinia vaccines that produce a "take" does not require use of the live variola virus. Use of the live virus will be necessary, however, for the development and licensure of any vaccine that does not manifest such a cutaneous lesion at the site of inoculation.

Development of Methods for Detection and Diagnosis

Contemporary nucleic acid-based methods for viral detection have been shown to identify variola virus genes directly, and multiplex PCR assays differentiate variola from other poxviruses and unrelated viruses, such as varicella-zoster virus, that may cause similar clinical signs. Since tissues contain inhibitors that may reduce the sensitivity and specificity of nucleic acid-based methods, the development of these assays is enhanced by the availability of stored clinical materials and specimens from nonhuman primates infected with variola virus, but these materials and specimens are not essential. Whether these methods have been tested with a representative set of phylogenetically and genetically diverse smallpox isolates is an important question, but further testing, if needed, does not require the growth of the live virus from existing stocks.

Protein-based assays have not been pursued as extensively as PCR methods; however, these methods can be tested using variola proteins made in expression vectors. Limited information has been published about the performance of any methods for environmental sampling to detect variola, but again such assessments do not require live variola virus. Licensing of these methods can also proceed without experiments using live virus. The primary barrier to development of these methods is a lack of development incentives and of a market for products that would allow rapid field detection and diagnosis.

> *The committee concludes that live variola virus is not required for further development of detection and diagnostic methods. Virus materials such as DNA and proteins would suffice for this purpose.*

Genomic Analysis

The past decade has seen advances in genome sequencing and functional genomics capabilities. As a result, significant progress has been made in acquiring new variola genome sequence data and in furthering understanding of the evolution of variola. This work has revealed significant sequence variability among variola strains, some of which is likely to be associated with virulence. The observed genetic differences between variola and other orthopoxviruses must be responsible for the specificity of variola virus for the human host. Variola genomic sequence data may enhance efforts to develop therapeutics and vaccines that are predicted to be active against the breadth of available variola strains. Despite the progress in sequencing variola strains, much remains to be learned about the extent of variola's genetic variability. In addition, the biological consequences of sequence differences for replication in particular cell types important to pathogenesis and to host range specificity and virulence are unknown. Today, sequencing of the genomes of all remaining variola strains to completion would be relatively straightforward and inexpensive.

The committee concludes that live variola virus is not needed for variola genome sequence analysis, as long as specimens containing viral DNA of adequate quantity and quality are available. Live variola virus would be needed for functional genomics-based experimental approaches.

Discovery Research

Variola virus can be useful for understanding human physiology and immunology because it has the capacity to overwhelm the host in a way that few viral pathogens do. Through studies in nonhuman primates, some progress has been made in understanding how variola virus modulates the functions of host cells for its benefit and how infection with the virus progresses in the host. However, current methods for studying variola in vitro and in vivo are inadequate or have not been fully exploited for the expeditious discovery of novel interventions, both for smallpox and for other diseases, that might result from a better understanding of how this pathogen takes over human cells and subverts the immune response. Further research is needed to develop improved animal models that can recapitulate key aspects of the human disease and to understand virus–cell interactions in human target cells relevant to pathogenesis and immune response.

The committee concludes that discovery research to gain greater understanding of human physiology and immunology, while not essential, would require use of the live variola virus and might ultimately support efforts to discover and evaluate therapeutics and vaccines. Further, research with live variola virus and research with variola proteins could lead to discoveries with broader implications for human health.

RECOMMENDATIONS

Gaps remain in understanding of variola virus and its interaction with its human host that could be critical in identifying potential targets for the discovery of therapeutics and vaccines. In particular, better understanding of the diversity and variability of variola strains would result in more effective therapeutics and vaccines, as well as more refined diagnostics. Genome sequencing could close existing knowledge gaps by illuminating differences among strains in molecular mechanisms of infection and response.

The committee recommends that WHO authorize the complete genome sequencing of all remaining variola strains, with the aim of understanding the patterns and extent of sequence variation and the relationships of these patterns to disease severity. This activity would be carried out at CDC, and ideally at VECTOR as well.

Similarly, a better understanding of variola pathogenesis would enhance the development of therapeutics and vaccines. Because smallpox is no longer naturally occurring, the closest approximation to human infection would involve a nonhuman primate. A more precise nonhuman primate model is essential for correct characterization of the efficacy of new therapeutics and vaccines. It is important to optimize approaches to infecting nonhuman primates so as to best recapitulate variola pathogenesis as it occurs in the human host, for example, by testing aerosol or intratracheal delivery as well as intravenous inoculation.

The committee recommends that a comprehensive evaluation of the work done to date on the nonhuman primate model of variola pathogenesis be undertaken by CDC, in conjunction with an expert panel knowledgeable about poxviruses and animal models of viral infection. The objective would be to identify ways in which the predictive value of the model for testing therapeutics and vaccines might be improved.

Finally, functional genomics tools, which are used to evaluate interactions between a replicating virus and the host cell, should be applied using a few representative variola strains in a number of representative differentiated human cell types. The purpose of this research would be to identify novel targets for therapeutics and to design third-generation vaccines.

The committee recommends that WHO explore the use of functional genomics approaches to improve understanding of variola pathogenesis and advance the development of novel strategies for therapeutic intervention.

Appendix

Variola Strains Used to Validate Diagnostic and Detection Assays

(Table on following pages.)

VAR Strain (Origin)	Method for Detection of Viral DNA	Probes/Targets	VAR from Other OPVs? (Y/N)
6/58, S/6	Microarray	*VAR C33L/B29R, B19R* genes[a]	Y
	Microarray	CrmB gene[b]	Y
12/62, S/12 (Tanzania)	Microarray	VAR C33L/B29R, B19R genes[a]	Y
	MPCR[c]	Five pairs of primers[d]	Y
	Microarray	CrmB gene[b]	Y
22/62, S/22	Microarray	VAR C33L/B29R, B19R genes[a]	Y
102, SAF65-102 (Republic of South Africa)	Real-Time PCR	VAR HA gene[e]	N
	Real-Time PCR	14-kDa protein gene[f]	Y
103, SAF65-103, Tvaal, Nelspruit (Republic of South Africa)	Real-Time PCR	VAR HA gene[e]	N
	Sequencing GeneChips	whole-genome sequences of 24 VAR strains[g]	N
	Real-Time PCR	14-kDa protein gene[f]	Y
7124 (India)	Real-Time PCR	VAR HA gene[e]	N
	Real-Time PCR	14-kDa protein gene[b]	Y
	Real-Time PCR	14-kDa protein gene[f]	Y
7125 (India)	Real-Time PCR	VAR HA gene[e]	N
	Real-Time PCR	14-kDa protein gene[f]	Y
66-39, CDC v66-39, Sao (Brazil)	Real-Time PCR	VAR HA gene[e]	N
	Sequencing GeneChips	whole-genome sequences of 24 VAR strains[g]	N
	Real-Time PCR	14-kDa protein gene[b]	Y
	Real-Time PCR	14-kDa protein gene[f]	Y
68-59 (Benin)	Real-Time PCR	VAR HA gene[e]	N
	Real-Time PCR	14-kDa protein gene[f]	Y

68-258 (Sierra Leone)	Real-Time PCR	VAR HA gene[e]	N
	Real-Time PCR	14-kDa protein gene[b]	Y
	Real-Time PCR	14-kDa protein gene[f]	Y
	PCR, RFLP	CrmB gene[i]	Y
70-222 (Indonesia)	Real-Time PCR	VAR HA gene[e]	N
	Real-Time PCR	14-kDa protein gene[f]	Y
70-228 (Indonesia)	Real-Time PCR	VAR HA gene[e]	N
		14-kDa protein gene[f]	Y
72-119 (Syria)	Real-Time PCR	VAR HA gene[e]	N
	Real-Time PCR	14-kDa protein gene[f]	Y
72-143 (Botswana)	Real-Time PCR	VAR HA gene[e]	N
	Real-Time PCR	14-kDa protein gene[f]	Y
73-175 (Nepal)	Real-Time PCR	VAR HA gene[e]	N
	Real-Time PCR	*pan*-E9L3, VAR B9R, VAR 10R genes[j]	N
73-225, BOT73_225, CDC v73-225 (Botswana)	Real-Time PCR	VAR HA gene[e]	N
	Sequencing GeneChips	whole-genome sequences of 24 VAR strains[g]	N
	Real-Time PCR	14-kDa protein gene[f]	Y
77-1252 (Somalia)	Real-Time PCR	VAR HA gene[e]	N
	Real-Time PCR	14-kDa protein gene[f]	Y
77-1605 (Somalia)	Real-Time PCR	VAR HA gene[e]	N
	Real-Time PCR	*pan*-E9L3, VAR B9R, VAR B10R genes[j]	Y

continued

VAR Strain (Origin)	Method for Detection of Viral DNA	Probes/Targets	VAR from Other OPVs? (Y/N)
77-227	Real-Time PCR	VAR HA gene[e]	N
Variolator 4 (Afghanistan)	Real-Time PCR	VAR HA gene[e]	N
	Real-Time PCR	14-kDa protein gene[b]	Y
	Real-Time PCR	14-kDa protein gene[f]	Y
	Real-Time PCR	*pan*-E9L3, VAR B9R, B10R genes[j]	Y
Aslam, S/ASL (Pakistan)	Microarray	VAR C33L/B29R, B19R genes[a]	Y
	MPCR[c]	five pairs of primers[d]	Y
	Microarray	CrmB gene[b]	Y
Aziz, S/AZI	Microarray	VAR C33L/B29R, B19R genes[a]	Y
	Microarray	CrmB gene[b]	Y
Variola major-ABA (Bangladesh)	Real-Time PCR	VAR HA gene[k]	N
Bangladesh-1975 (Bangladesh)	Microarray	VAR C33L/B29R, B19R genes[a]	Y
	Real-Time PCR	14-kDa protein gene[f]	Y
	PCR, RFLP	CrmB gene[i]	Y
Bombay (India)	Real-Time PCR	VAR HA gene[e]	N
Brazil 128, S/B128 (Brazil)	Microarray	VAR C33L/B29R, B19R genes[a]	Y
	MPCR[c]	five pairs of primers[d]	Y
Brazil 131, S/B131 (Brazil)	Microarray	VAR C33L/B29R, B19R genes[a]	Y
	MPCR[c]	five pairs of primers[d]	Y

Brazil Garcia (Brazil)	Real-Time PCR	VAR HA gene[e]	N
	Real-Time PCR	14-kDa protein gene[b]	Y
	Microarray	VAR C33L/B29R, B19R genes[a]	Y
	Real-Time PCR	14-kDa protein gene[f]	Y
	Microarray	CrmB gene[b]	Y
	PCR, RFLP	CrmB gene[i]	Y
BSH	Real-Time PCR	VAR HA gene[e]	N
BSH75_banu, CDC v75-550 (Bangladesh)	Sequencing GeneChips	whole-genome sequences of 24 VAR strains[g]	N
	Real-Time PCR	*pan*-E9L3, VAR B9R, VAR B10R genes[j]	Y
Butler, S/BUT (United Kingdom)	Real-Time PCR	VAR HA gene[e]	N
	Microarray	VAR C33L/B29R, B19R genes[a]	Y
	MPCR[c]	five pairs of primers[d]	Y
	Real-Time PCR	14-kDa protein gene[f]	Y
	Microarray	CrmB gene[b]	Y
China Horn, CHN48_horn (China)	Sequencing GeneChips	whole-genome sequences of 24 VAR strains[h]	N
	Real-Time PCR	*pan*-E9L3, VAR B9R, VAR B10R genes[j]	Y
Congo (Congo)	Real-Time PCR	VAR HA gene[e]	N
	Real-Time PCR	14-kDa protein gene[f]	Y
	PCR, RFLP	CrmB gene[i]	Y
v70-46 (Congo)	Real-Time PCR	*pan*-E9L3, VAR B9R, VAR B10R genes[j]	Y
Variola major-CNG (Congo)	Real-Time PCR	VAR HA gene[k]	N
Congo-2, S/CNG2 (Congo)	Microarray	VAR C33L/B29R, B19R genes[a]	Y
	MPCR[c]	five pairs of primers[d]	Y
	Microarray	CrmB gene[b]	Y

continued

VAR Strain (Origin)	Method for Detection of Viral DNA	Probes/Targets	VAR from Other OPVs? (Y/N)
	Microarray	C23L/B29R gene[l]	Y
Congo-9, v74-227, S/CNG9 (Congo)	Microarray	VAR C33L/B29R, B19R genes[a]	Y
	Real-Time PCR	14-kDa protein gene[b]	Y
	MPCR[c]	five pairs of primers[d]	Y
	Microarray	CrmB gene[b]	Y
Ethiopia 16, ETH72-16, R14-IX-72, Addis (Ethiopia)	Sequencing GeneChips	whole-genome sequences of 24 VAR strains[g]	N
	Real-Time PCR	14-kDa protein gene[f]	Y
Ethiopia 17, ETH72-17 (Ethiopia)	Real-Time PCR	VAR HA gene[e]	N
	Sequencing GeneChips	whole-genome sequences of 24 VAR strains[g]	N
	Real-Time PCR	14-kDa protein gene[f]	Y
Variola minor-GAR (Garcia)	Real-Time PCR	VAR HA gene[k]	N
Variola major-HAR	Real-Time PCR	VAR HA gene[k]	N
Harper (Japan)	Real-Time PCR	VAR HA gene[e]	N
	Real-Time PCR	14-kDa protein gene[f]	Y
Harvey (United Kingdom)	Real-Time PCR	VAR HA gene[e]	N
	Real-Time PCR	14-kDa protein gene[f]	Y
	PCR, RFLP	CrmB gene[i]	Y
Heidelberg, GER_hdlg (Germany, Asia import)	Real-Time PCR	VAR HA gene[e]	N
	Sequencing GeneChips	whole-genome sequences of 24 VAR strains[g]	N
	Real-Time PCR	14-kDa protein gene[b]	Y
	Real-Time PCR	14-kDa protein gene[f]	Y

Herrlich (India)	Real-Time PCR	14-kDa protein gene[f]	Y
Hembula (Tanzania)	Real-Time PCR	14-kDa protein gene[f]	Y
Higgins, UNK47_hig (United Kingdom)	Real-Time PCR	VAR HA gene[e]	N
	Sequencing GeneChips	whole-genome sequences of 24 VAR strains[g]	N
	Real-Time PCR	14-kDa protein gene[h]	Y
	Real-Time PCR	14-kDa protein gene[h]	Y
Helder, S/HEL	Microarray	VAR C33L/B29R, B19R genes[a]	Y
Hinden (United Kingdom)	Real-Time PCR	14-kDa protein gene[f]	Y
Hinton	Real-Time PCR	VAR HA gene[e]	N
Horn (China)	Real-Time PCR	VAR HA gene[e]	N
	Real-Time PCR	14-kDa protein gene[f]	Y
Ind-3a	Microarray	VAR C33L/B29R, B19R genes[a]	Y
	MPCR[c]	five pairs of primers[d]	Y
	Microarray	CrmB gene[b]	Y
	Microarray	C23L/B29R gene[l]	Y
Variola major-BOM (India)	Real-Time PCR	VAR HA gene[k]	N
Variola major-MAD (India)	Real-Time PCR	VAR HA gene[k]	N
India, S/IND (India)	Microarray	VAR C33L/B29R, B19R genes[a]	Y

continued

VAR Strain (Origin)	Method for Detection of Viral DNA	Probes/Targets	VAR from Other OPVs? (Y/N)
India 164, S/I164 (India)	Microarray	VAR C33L/B29R, B19R[a]	Y
	MPCR[c]	five pairs of primers[d]	Y
India 378 (India)	Microarray	VAR C33L/B29R, B19R[a]	Y
	Microarray	CrmB gene[b]	Y
India-1967, S/IND67 (India)	Microarray	VAR C33L/B29R, B19R[a]	Y
India 7124 (India)	Real-Time PCR	VAR HA, B9R, B10R genes[j]	Y
India 7125 (India)	Real-Time PCR	14-kDa protein gene[h]	Y
Iran (Iran)	Real-Time PCR	VAR HA gene[e]	N
IRN72_tbrz, Iran 2602, Tabrez (Iran)	Sequencing GeneChips	whole-genome sequences of 24 VAR strains[g]	N
	Real-Time PCR	14-kDa protein gene[f]	Y
Juba (Sudan)	Real-Time PCR	VAR HA gene[e]	N
	Real-Time PCR	14-kDa protein gene[f]	Y
K1629 (Kuwait)	Real-Time PCR	VAR HA gene[e]	N
	Real-Time PCR	14-kDa protein gene[f]	Y
Kali Mathu (India)	Real-Time PCR	VAR HA gene[e]	N
	Real-Time PCR	14-kDa protein gene[f]	Y
Kembula	Real-Time PCR	VAR HA gene[e]	N
	Real-Time PCR	14-kDa protein gene[h]	Y

Khateen, S/WSI (Pakistan)	Microarray	VAR C33L/B29R, B19R genes[a]	Y
	MPCR[c]	five pairs of primers[d]	Y
Kudano (Nigeria)	Real-Time PCR	14-kDa protein gene[f]	Y
Kuw-5, S/DUW5 (Kuwait)	Microarray	VAR C33L/B29R, B19R genes[a]	Y
	MPCR[c]	five pairs of primers[d]	Y
	Microarray	CrmB gene[b]	Y
	Microarray	C23L/B29R gene[l]	Y
Lee (Korea)	Real-Time PCR	14-kDa protein gene[f]	Y
M-Abr-60, S/ABR (Russia)	Microarray	VAR C33L/B29R, B19R genes[a]	Y
	MPCR[c]	five pairs of primers[d]	Y
	Microarray	CrmB gene[b]	Y
M-Sok-60, S/SOK (Russia)	Microarray	VAR C33L/B29R, B19R genes[a]	Y
	MPCR[c]	five pairs of primers[d]	Y
	Microarray	CrmB gene[b]	Y
M-Sur-60, S/SUR (Russia)	Microarray	VAR C33L/B29R, B19R genes[a]	Y
	MPCR[c]	five pairs of primers[d]	Y
	Microarray	CrmB gene[b]	Y
M-N-60	Microarray	VAR C33L/B29R, B19R genes[a]	Y
Mannan (Bangladesh)	Real-Time PCR	14-kDa protein gene[f]	Y
Minnesota124 (United States)	Real-Time PCR	VAR HA gene[e]	N
	Real-Time PCR	14-kDa protein gene[f]	Y
MS Lee	Real-Time PCR	VAR HA gene[e]	N
	Real-Time PCR	14-kDa protein gene[h]	Y

continued

VAR Strain (Origin)	Method for Detection of Viral DNA	Probes/Targets	VAR from Other OPVs? (Y/N)
Nepal (Nepal)	Real-Time PCR	VAR HA gene[e]	N
Nepal 73 (Nepal)	Real-Time PCR	14-kDa protein gene[f]	Y
New Delhi, IND53_ndel (India)	Real-Time PCR	VAR HA gene[e]	N
	Sequencing GeneChips	whole-genome sequences of 24 VAR strains[g]	N
	Real-Time PCR	14-kDa protein gene[f]	Y
Ngami, S/NGA (Tanzania)	Microarray	VAR C33L/B29R, B19R genes[a]	Y
	MPCR[c]	five pairs of primers[d]	Y
	Microarray	CrmB gene[b]	Y
Variola major-NIG (Nigeria)	Real-Time PCR	VAR HA gene[k]	N
Nigeria Hudano (Nigeria)	Real-Time PCR	VAR HA gene[e]	N
Nur Islam, BSH74_nur (Bangladesh)	Real-Time PCR	VAR HA gene[e]	N
	Sequencing GeneChips	whole-genome sequences of 24 VAR strains[g]	N
	Real-Time PCR	14-kDa protein gene[f]	Y
Parvin (Bangladesh)	Real-Time PCR	14-kDa protein gene[f]	Y
Variola major-RAT (Pakistan)	Real-Time PCR	VAR HA gene[k]	N
Rafiq, Lahore, PAK69_lah (Pakistan)	Real-Time PCR	VAR HA gene[e]	N
	Sequencing GeneChips	whole-genome sequences of 24 VAR strains[g]	N
	Real-Time PCR	14-kDa protein gene[f]	Y
Rumbec (Sudan)	Real-Time PCR	VAR HA gene[e]	N
	Real-Time PCR	14-kDa protein gene[f]	Y

Rw18, S/RW18	Microarray	VAR C33L/B29R, B19R genes[a]	Y
Semat, S/SEM (Tanzania)	Microarray	VAR C33L/B29R, B19R genes[a]	Y
	MPCR[c]	five pairs of primers[d]	Y
	Microarray	CrmB gene[b]	Y
Shahzaman (Bangladesh)	Real-Time PCR	VAR HA gene[e]	N
	Real-Time PCR	14-kDa protein gene[h]	Y
	Real-Time PCR	14-kDa protein gene[f]	Y
Variola major-SLN (Sierra Leone)	Real-Time PCR	VAR HA gene[k]	N
Solaiman (Bangladesh)	Real-Time PCR	VAR HA gene[e]	N
	Real-Time PCR	14-kDa protein gene[f]	Y
SOM77_ali (Somalia-1977, CDC V77-2479 [last case])	Sequencing GeneChips	whole-genome sequences of 24 VAR strains[g]	N
	Microarray	VAR C33L/B29R, B19R genes[a]	Y
	Real-Time PCR	VAR HA gene[k]	N
	PCR, RFLP	CrmB gene[i]	Y
Stillwell (Japan)	Real-Time PCR	VAR HA gene[e]	N
	Real-Time PCR	14-kDa protein gene[f]	Y
Taj Barin, S/TAJ	Microarray	VAR C33L/B29R, B19R genes[a]	Y
Variola minr-BUT (United Kingdom)	Real-Time PCR	VAR HA gene[k]	N
Yamada (Japan)	Real-Time PCR	VAR HA gene[e]	N
	Real-Time PCR	14-kDa protein gene[f]	Y

continued (notes on following page)

NOTES:
[a]Ryabinin et al., 2006.
[b]Lapa et al., 2002. The microarray contained 15 oligonucleotide probes directed toward five species-specific segments of the CrmB gene.
[c]MPCR = Multiplex PCR.
[d]Schelkunov et al., 2005. Five pairs of oligonucleotide primers—one genus-specific, and the rest species-specific for variola, monkeypox, cowpox, and vaccinia viruses.
[e]Ibrahim et al., 2003.
[f]Olson et al., 2004.
[g]Sulaiman et al., 2007. Seven GeneChips, containing ~240,000 different types of 25-mer oligonucleotides, each designed to analyze a divergent segment of approximately 30,000 bases of VAR genome.
[h]Scaramozzino et al., 2007. One probe to recognize human pathogenic orthopoxviruses (OPV) and one specific for VAR.
[i]Loparev et al., 2001.
[j]Kulesh et al., 2004. Assays used cloned genes.
[k]Schoepp et al., 2004.
[l]Laassri et al., 2003.

REFERENCES

Ibrahim, M. S., D. A. Kulesh, S. S. Saleh, I. K. Damon, J. J. Esposito, A. L. Schmaljohn, and P. B. Jahrling. 2003. Real-time PCR assay to detect smallpox virus. *Journal of Clinical Microbiology* 41(8):3835–3839.

Kulesh, D. A., R. O Baker, B. M. Loveless., D. Norwood, S. H., Zwiers, E. Mucker, C. Hartmann, R. Herrera, D. Miller, D. Christensen, L.P., Wasieloski, J. Huggins, and P. B. Jahrling. 2004. Smallpox and pan-orthopox virus detection by real-time 3′-minor groove binder TaqMan assays on the roche LightCycler and the Cepheid smart Cycler platforms. *Journal of Clinical Microbiology* 42(2):601–609.

Laassri, M., V. Chizhikov, M. Mikheev, S. Shchelkunov, and K. Chumakov. 2003. Detection and discrimination of orthopoxviruses using microarrays of immobilized oligonucleotides. *Journal of Virological Methods* 112(1–2):67–78.

Lapa, S., M. Mikheev, S. Shchelkunov, V. Mikhailovich, A. Sobolev, V. Blinov, L. Babkin, A. Guskov, E. Sokunova, A. Zasedatelev, L. Sandakhchiev, and A. Mirzabekov. 2002. Species-level identification of orthopoxviruses with an oligonucleotide microchip. *Journal of Clinical Microbiology* 40(3):753–757.

Loparev, V. N., R. F. Massung, J. J. Esposito, and H. Meyer. 2001. Detection and differentiation of Old World orthopoxviruses: Restriction fragment length polymorphism of the crmB gene region. *Journal of Clinical Microbiology* 39(1):94–100.

Olson, V. A., T. Laue, M. T. Laker, I. V. Babkin, C. Drosten, S. N. Shchelkunov, M. Niedrig, I. K. Damon, and H. Meyer. 2004. Real-Time PCR System for Detection of Orthopoxviruses and Simultaneous Identification of Smallpox Virus. *Journal of Clinical Microbiology* 42(5):1940–1946.

Ryabinin, V. A., Shundrin, L.A., Kostina, E.B., Laassri, M., Chizhikov, V., Shchelkunov, S.N., Chumakov, K., and A. N. Sinyakov. 2006. Microarray assay for detection and discrimination of Orthopoxvirus species. *Journal of Medical Virology* 78(10):1325–1340.

Scaramozzino, N., A. Ferrier-Rembert, A.L. Favier, C. Rothlisberger, S. Richard, J.M. Crance, H. Meyer, and D. Garin. 2007. Real-time PCR to identify variola virus or other human pathogenic orthopox viruses. *Clinical Chemistry* 53(4):606–613.

Schoepp, R. J., M.D. Morin, M.J., Martinez, D. A. Kulesh, L. Hensley, T. W. Geisbert, D. R. Brady, and P. B. Jahrling. 2004. Detection and identification of Variola virus in fixed human tissue after prolonged archival storage. *Laboratory Investigation* 84(1):41–48.

Shchelkunov, S. N., E. V. Gavrilova, and I. V. Babkin. 2005. Multiplex PCR detection and species differentiation of orthopoxviruses pathogenic to humans. *Molecular & Cellular Probes* 19(1):1–8.

Sulaiman, I.M., T. Tang, J. Osborne, S. Sammons, and R. M. Wohlhueter. 2007. GeneChip resequencing of the smallpox virus genome can identify novel strains: A biodefense application. *Journal of Clinical Microbiology* 45(2):358–363.